Assessment of Mission Size Trade-offs for NASA's Earth and Space Science Missions

Ad Hoc Committee on the Assessment of Mission Size Trade-offs for
Earth and Space Science Missions

Space Studies Board

Commission on Physical Sciences, Mathematics, and Applications

National Research Council

NATIONAL ACADEMY PRESS
Washington, D.C.

NOTICE: The project that is the subject of this report was approved by the Governing Board of the National Research Council, whose members are drawn from the councils of the National Academy of Sciences, the National Academy of Engineering, and the Institute of Medicine. The members of the committee responsible for the report were chosen for their special competences and with regard for appropriate balance.

Support for this project was provided by Contract NASW 96013 between the National Academy of Sciences and the National Aeronautics and Space Administration. Any opinions, findings, conclusions, or recommendations expressed in this material are those of the authors and do not necessarily reflect the views of the sponsor.

International Standard Book Number 0-309-06976-9

Copyright 2000 by the National Academy of Sciences. All rights reserved.

Copies of this report are available free of charge from:

Space Studies Board
National Research Council
2101 Constitution Avenue, N.W.
Washington, D.C. 20418

Printed in the United States of America

THE NATIONAL ACADEMIES

National Academy of Sciences
National Academy of Engineering
Institute of Medicine
National Research Council

The **National Academy of Sciences** is a private, nonprofit, self-perpetuating society of distinguished scholars engaged in scientific and engineering research, dedicated to the furtherance of science and technology and to their use for the general welfare. Upon the authority of the charter granted to it by the Congress in 1863, the Academy has a mandate that requires it to advise the federal government on scientific and technical matters. Dr. Bruce M. Alberts is president of the National Academy of Sciences.

The **National Academy of Engineering** was established in 1964, under the charter of the National Academy of Sciences, as a parallel organization of outstanding engineers. It is autonomous in its administration and in the selection of its members, sharing with the National Academy of Sciences the responsibility for advising the federal government. The National Academy of Engineering also sponsors engineering programs aimed at meeting national needs, encourages education and research, and recognizes the superior achievements of engineers. Dr. William A. Wulf is president of the National Academy of Engineering.

The **Institute of Medicine** was established in 1970 by the National Academy of Sciences to secure the services of eminent members of appropriate professions in the examination of policy matters pertaining to the health of the public. The Institute acts under the responsibility given to the National Academy of Sciences by its congressional charter to be an adviser to the federal government and, upon its own initiative, to identify issues of medical care, research, and education. Dr. Kenneth I. Shine is president of the Institute of Medicine.

The **National Research Council** was organized by the National Academy of Sciences in 1916 to associate the broad community of science and technology with the Academy's purposes of furthering knowledge and advising the federal government. Functioning in accordance with general policies determined by the Academy, the Council has become the principal operating agency of both the National Academy of Sciences and the National Academy of Engineering in providing services to the government, the public, and the scientific and engineering communities. The Council is administered jointly by both Academies and the Institute of Medicine. Dr. Bruce M. Alberts and Dr. William A. Wulf are chairman and vice chairman, respectively, of the National Research Council.

AD HOC COMMITTEE ON THE ASSESSMENT OF MISSION SIZE TRADE-OFFS FOR EARTH AND SPACE SCIENCE MISSIONS

DANIEL N. BAKER, University of Colorado, *Chair*
FRAN BAGENAL, University of Colorado
ROBERT L. CAROVILLANO, Boston College
RICHARD G. KRON, University of Chicago
GEORGE A. PAULIKAS, The Aerospace Corporation (retired)
R. KEITH RANEY, Johns Hopkins University
PEDRO L. RUSTAN, JR., U.S. Air Force (retired)

Discipline Committee Liaisons

WENDY CALVIN, U.S. Geological Survey (Committee on Planetary and Lunar Exploration)
ROBERT L. CAROVILLANO, Boston College (Committee on Solar and Space Physics)
RICHARD F. MUSHOTZKY, NASA Goddard Space Flight Center (Committee on Astronomy and Astrophysics (CAA))
DEANE PETERSON, State University of New York (CAA)
BLAIR D. SAVAGE, Washburn Observatory (CAA)
LAWRENCE C. SCHOLZ, West Orange, New Jersey (Committee on Earth Studies)

Staff

PAMELA L. WHITNEY, Study Director
TAMARA DICKINSON, Senior Program Officer
REBECCA SHAPACK, Research Assistant
CARMELA CHAMBERLAIN, Senior Program Assistant

SPACE STUDIES BOARD

CLAUDE R. CANIZARES, Massachusetts Institute of Technology, *Chair*
MARK R. ABBOTT, Oregon State University
FRAN BAGENAL, University of Colorado
DANIEL N. BAKER, University of Colorado
ROBERT E. CLELAND, University of Washington
MARILYN L. FOGEL, Carnegie Institution of Washington
BILL GREEN, former member, U.S. House of Representatives
JOHN H. HOPPS, JR., Morehouse College
CHRISTIAN J. JOHANNSEN, Purdue University
RICHARD G. KRON, University of Chicago
JONATHAN I. LUNINE, University of Arizona
ROBERTA BALSTAD MILLER, Columbia University
GARY J. OLSEN, University of Illinois at Urbana-Champaign
MARY JANE OSBORN, University of Connecticut Health Center
GEORGE A. PAULIKAS, The Aerospace Corporation (retired)
JOYCE E. PENNER, University of Michigan
THOMAS A. PRINCE, California Institute of Technology
PEDRO L. RUSTAN, JR., U.S. Air Force (retired)
GEORGE L. SISCOE, Boston University
EUGENE B. SKOLNIKOFF, Massachusetts Institute of Technology
MITCHELL SOGIN, Marine Biological Laboratory
NORMAN E. THAGARD, Florida State University
ALAN M. TITLE, Lockheed Martin Advanced Technology Center
RAYMOND VISKANTA, Purdue University
PETER W. VOORHEES, Northwestern University
JOHN A. WOOD, Harvard-Smithsonian Center for Astrophysics

JOSEPH K. ALEXANDER, Director

COMMISSION ON PHYSICAL SCIENCES, MATHEMATICS, AND APPLICATIONS

PETER M. BANKS, Veridian ERIM International, Inc., *Co-chair*
W. CARL LINEBERGER, University of Colorado, *Co-chair*
WILLIAM F. BALLHAUS, JR., Lockheed Martin Corporation
SHIRLEY CHIANG, University of California at Davis
MARSHALL H. COHEN, California Institute of Technology
RONALD G. DOUGLAS, Texas A&M University
SAMUEL H. FULLER, Analog Devices, Inc.
JERRY P. GOLLUB, Haverford College
MICHAEL F. GOODCHILD, University of California at Santa Barbara
MARTHA P. HAYNES, Cornell University
WESLEY T. HUNTRESS, JR., Carnegie Institution
CAROL M. JANTZEN, Westinghouse Savannah River Company
PAUL G. KAMINSKI, Technovation, Inc.
KENNETH H. KELLER, University of Minnesota
JOHN R. KREICK, Sanders, a Lockheed Martin Company (retired)
MARSHA I. LESTER, University of Pennsylvania
DUSA M. McDUFF, State University of New York at Stony Brook
JANET L. NORWOOD, Former Commissioner, U.S. Bureau of Labor Statistics
M. ELISABETH PATÉ-CORNELL, Stanford University
NICHOLAS P. SAMIOS, Brookhaven National Laboratory
ROBERT J. SPINRAD, Xerox PARC (retired)

MYRON F. UMAN, Acting Executive Director

Preface

In the mid-1990s, NASA began to reorient its approach to space science missions by placing increased emphasis on ways to make projects faster, better, and cheaper. The faster-better-cheaper (FBC) label generally refers to space research missions such as those in the small and medium Explorer series and the Discovery and Earth System Science Pathfinder lines. These missions are allotted 3 or 4 years for completion from the time they are selected. Costs range from less than $150 million to approximately $350 million.[1] The emphasis NASA is placing on faster-better-cheaper missions has created the impression that it may have completely abandoned the larger missions it had been known for in the past. Concerned about this impression, Congress directed NASA to "contract with the National Research Council (NRC) for a study across all space science and Earth science disciplines to identify missions that cannot be accomplished within the parameters imposed by the smaller, faster, cheaper, better regime" (see Appendix A). This report represents the response of the National Research Council's (NRC's) Space Studies Board (SSB) to that congressional request. Based on understanding of the information needed, three tasks were formulated for the ad hoc committee conducting the study (Appendix B).

As the SSB noted in its approach to this assessment, NASA's FBC strategy has involved efforts to streamline mission development cycles, thereby increasing the number and frequency of flight missions. In principle, this should provide more opportunities for investigators to access spaceflight data. Such missions can be developed and launched in a few years, at a flight rate of 10 or more per year and at costs of no more than a few hundred million dollars each. In contrast, traditional large missions such as Viking, Voyager, Galileo, the Hubble Space Telescope, the Upper Atmosphere Research Satellite (UARS), and the Earth Observing System (EOS) Terra mission have each required a decade or more to develop and budgets from several hundred million to several billion dollars. However, it is also true that FBC missions can involve certain scientific sacrifices and risks: for example, when they require compromises in the breadth or depth of the measurements that can be accomplished or when design practices or technology features require risk-taking to meet cost constraints.

The approach to conducting the study was determined by the Space Studies Board at its meeting on June 22-24, 1999, at the NASA John Glenn Research Center. The board decided to assemble an ad hoc committee comprising a subset of board members with expertise in the Earth sciences, astronomy and astrophysics, space

---

[1] For the purposes of this study, NASA defined "small" as missions with total life-cycle costs of less than $150 million, "medium" as between $150 million and $350 million, and "large" as more than $350 million.

physics, planetary sciences, and space technology to conduct the study. The committee worked with liaison members from four of the board's discipline committees and also received input from one of its interdisciplinary committees.[2] These committees were assigned a series of questions (see Appendix C) and asked to provide written materials for the ad hoc committee. Their contributions (Appendix E) provided the raw material for the report. The Ad Hoc Committee on the Assessment of Mission Size Trade-offs for Earth and Space Science Missions met concurrently with the Space Studies Board Executive Committee in Woods Hole, Massachusetts, on September 8-10, 1999, and again on November 10-11, 1999, at the NASA Stennis Space Center (Appendix D).

Chapter 1 of the committee's report outlines the central issues and considerations for assessing mission size trade-offs for Earth and space science programs, including (1) fundamental science limits, (2) scientific return, (3) mission implementation, (4) technology, (5) access to space, (6) risk, and (7) problems with past missions. Chapter 2 identifies and illustrates arguments for evaluating a portfolio of mission sizes in the various sub-disciplines of Earth and space sciences. Chapter 3 revisits the tasks assigned in the charge and presents the committee's findings and recommendations.

---

[2]The Committee on Astronomy and Astrophysics (CAA), the Committee on Earth Studies (CES), the Committee on Planetary and Lunar Exploration (COMPLEX), the Committee on Solar and Space Physics (CSSP), and the Committee on International Space Programs (CISP).

Acknowledgments

This report has been reviewed by individuals chosen for their diverse perspectives and technical expertise, in accordance with procedures approved by the National Research Council's (NRC's) Report Review Committee. The purpose of this independent review is to provide candid and critical comments that will assist the authors and the NRC in making the published report as sound as possible and to ensure that the report meets institutional standards for objectivity, evidence, and responsiveness to the study charge. The contents of the review comments and draft manuscript remain confidential to protect the integrity of the deliberative process. The committee wishes to thank the following individuals for their participation in the review of this report:

Peter Burr, NASA Goddard Space Flight Center (retired),
John R. Casani, Jet Propulsion Laboratory (retired),
Marshall H. Cohen, California Institute of Technology,
Richard Goody, Harvard University (emeritus),
Marcia J. Rieki, University of Arizona,
Byron Tapley, University of Texas,
Joseph Veverka, Cornell University, and
Donald Williams, Johns Hopkins University.

Although the individuals listed above have provided many constructive comments and suggestions, responsibility for the final content of this report rests solely with the authoring committee and the NRC.

The committee also wishes to acknowledge Kenneth Ledbetter, Mission and Payload Development Division, Office of Space Science, NASA, and his staff, and Andrew Hunter, Business Division, Office of Earth Science, NASA, for providing information on the costs of missions.

Contents

EXECUTIVE SUMMARY 1

1 ISSUES AND CONSIDERATIONS IN THE ASSESSMENT OF MISSION SIZE
   TRADE-OFFS IN THE EARTH AND SPACE SCIENCES 6
   Fundamental Science Limits, 7
   Measuring and Enhancing the Scientific Return on Investment, 14
   Implementation, 16
   Technology, 24
   Access to Space, 27
   Risk, 27
   Problems with Past Missions, 28

2 SCIENCE PRIORITIES AND NASA MISSION PLANS 31
   Introduction, 31
   Cross-Cutting Themes, 32
   Discipline-Specific Issues and Concerns, 38

3 SUMMARY AND RECOMMENDATIONS 52
   The Charge, 52
   Strengths and Weaknesses of Small and Large Missions, 53
   Recommendations in Response to the Charge, 54
   Other Findings on Issues Affecting Mission Size Mix, 56

APPENDIXES
A Letter of Request from NASA to the Space Studies Board 61
B Statement of Task 65
C Information Sought from Space Studies Board Discipline Committees 67
D Meeting Agenda 69
E Material Provided by Space Studies Board Discipline Committees 71
F Acronyms and Abbreviations 86
G Biographies of Committee Members 90

Executive Summary

This report addresses fundamental issues of mission architecture in the nation's scientific space program and responds to the FY99 Senate conference report,[1] which requested that NASA commission a study to assess the strengths and weaknesses of small, medium, and large missions. To that end, three tasks were set for the Ad Hoc Committee on the Assessment of Mission Size Trade-offs for Earth and Space Science Missions:

1. Evaluate the general strengths and weaknesses of small, medium, and large missions[2] in terms of their potential scientific productivity, responsiveness to evolving opportunities, ability to take advantage of technological progress, and other factors that may be identified during the study;
2. Identify which elements of the SSB and NASA science strategies will require medium or large missions to accomplish high-priority science objectives; and
3. Recommend general principles or criteria for evaluating the mix of mission sizes in Earth and space science programs. The factors to be considered will include not only scientific, technological, and cost trade-offs but also institutional and structural issues pertaining to the vigor of the research community, government-industry-university partnerships, graduate student training, and the like.

The committee approached these questions in light of the changing environment at NASA, which has been conducting an increasing number of smaller space and Earth science missions having shorter development times and using streamlined management methods, advanced technologies, and more compact platforms than had been employed in the past. The committee referred to this approach as the faster-better-cheaper (FBC) paradigm, a variant of "smaller, faster, cheaper, better" and similar phrases that have been used to describe the changing environment for space research missions.

The committee interpreted the FBC paradigm as a set of principles (including, but not limited to, streamlined management, flexibility, and technological capability) that are independent of the size or scope of a mission but

---

[1] U.S. Senate. 1998. Department of Veterans Affairs, Housing and Urban Development, and Independent Agencies Appropriations Bill, 1999, 105th Congress, 2nd Sess., S. Rept. 105-216.
[2] For the purposes of this study, NASA defined "small" as missions with total life-cycle costs less than $150 million, "medium" as between $150 million and $350 million, and "large" as more than $350 million.

can be matched appropriately to the science objectives and requirements for a given mission. It understood the term "mission" to mean the entire process of carrying out a space-based research activity, including scientific conception, spacecraft and instrument design and development, selection of development contractors, development costs, selection of launch capability, launch costs, mission operations, data analysis, and dissemination of scientific results.

It is within this broad context that the committee considered questions about the emerging FBC paradigm and its implications for mission size mixes in NASA's Earth and space science programs. How FBC is defined and how FBC principles are applied to programs of any scale have many implications for the space program: its tolerance for risk; its ability to carry out strategic plans; the scope, scale, and diversity of science investigated; the results and analytical products of its missions; the ways it trains young scientists and engineers; the role of international cooperation and the ease with which it can be incorporated into NASA's programs and plans; the role of universities, industry, government laboratories, and NASA centers in conducting space research missions; and the general health and vitality of the space science and Earth science enterprises. Policy makers looking for guidance on these programs in terms of cost and size trade-offs should be made aware that the variables are more numerous and much more complex than might at first be supposed.

The FBC approach emerged from the widely held belief that some large, traditional NASA missions had become unwieldy. With development times of over a decade (which often resulted in flying less capable technologies) and escalating costs, such missions came under increasing scrutiny, even given the magnificence of their promised (and realized) scientific returns. Traditional missions called into question the ability of NASA's Earth and space science research programs to obtain the highest quality and quantity of research return in the most timely and efficient fashion. Cuts in NASA's budget beginning in the early 1990s further encouraged new approaches for obtaining scientific returns in more efficient and cost-effective ways, albeit with added risk.

"Faster" missions can be made so by streamlining the management and development effort, by shortening the development schedule, by using the best available technology, and perhaps even by knowingly accepting more risk. In general, such methods will also lead to a "cheaper" mission. However, for NASA research programs, technological or managerial innovation are not ends unto themselves: the clear and obvious meaning of "better" is that more science—more knowledge and better quality and quantity of measurements—about some aspects of the universe around us is returned for a given investment and that such returns occur in a timely manner.

The impression that faster-better-cheaper also means "smaller" has raised concerns that there is a growing shift away from larger-scale endeavors in the Earth and space science programs. However, the tendency to equate FBC with the size or cost of a space or Earth science mission can overlook a number of things: the requirements unique to different disciplines, the complexities of scientific objectives, time and spatial scales, and techniques for implementing a mission. Total costs, mission capabilities, and the ultimate scientific results of space programs rely on a complex combination of the skill and performance of everyone associated with mission development, schedules, approaches to handling technical and management risks, technological implementation, and management style.

Through the careful planning processes that now characterize both the Earth science and the space science enterprises, the key outstanding questions of each discipline can be framed. Each such science question or disciplinary quest must then be examined in terms of the science community's priorities, the measurement requirements, and the technological readiness to determine which mission approach (or approaches) might be employed to address it. These science-based decisions on missions and approaches also incorporate strategies to engage and educate the general public and contribute to broader goals such as human exploration and development of space. A major consideration in all cases is the fiscal constraint that applies at any given time and the level of risk that can be tolerated by the mission's scientific priority and its role in NASA's strategic plan.

The ad hoc committee recognizes that the recent losses of missions conducted using the FBC approach—Lewis, the Wide-Field Infrared Explorer, Mars Climate Observer, and Mars Polar Lander—are in many ways calling into question some elements of the philosophy of FBC. Although it is beyond the scope of the committee's charge to assess individual mission failures (this is a task for the mission failure review boards), the committee calls attention to the potential implications of these losses for science and, especially, for the direction of the

NASA Mars program. Is the Mars program committed to a technology path that is proving to be riskier than its proponents originally anticipated? Are recent losses turning the program toward sample return missions that lack the critical precursors recommended in science strategy reports? How seriously have the scientific rationale and robustness of the Mars program been affected by the information lost from recent mission failures? Do current and future mission programs have ample time and budgets to integrate the lessons learned from previous failures? These and other ramifications of the recent series of losses of missions implemented under the FBC paradigm are of pressing and paramount concern.

FINDINGS

The committee supports several principles being implemented in the FBC methodology. Specifically, it found a number of positive aspects of the FBC approach, including the following:

- A mixed portfolio of mission sizes is crucial in virtually all Earth and space science disciplines to accomplish the various research objectives. The FBC approach has produced useful improvements across the spectrum of programs regardless of absolute mission size or cost.
- Shorter development cycles have enhanced scientific responsiveness, lowered costs, involved a larger community, and enabled the use of the best available technologies.
- The increased frequency of missions has broadened research opportunities for the Earth and space sciences.
- Scientific objectives can be met with greater flexibility by spreading a program over several missions.

Nonetheless, some problems exist in the practical application of the FBC approach, including the following:

- The heavy emphasis on cost and schedule has too often compromised scientific outcomes (scope of mission, data return, and analysis of results).
- Technology development is a cornerstone of the FBC approach for science missions but is often not aligned with science-based mission objectives.
- The cost and schedule constraints for some missions may lead to choosing designs, management practices, and technologies that introduce additional risks.
- The nation's launch infrastructure is limited in its ability to accommodate smaller spacecraft in a timely, reliable, and cost-effective way.

RECOMMENDATIONS TO NASA

Faster-Better-Cheaper Principles

Faster-better-cheaper methods of management, technology infusion, and implementation have produced useful improvements regardless of absolute mission size or cost. However, while improvements in administrative procedures have proven their worth in shortening the time to science, experience from mission losses (Mars Climate Observer and Lewis, for example) has shown that great care must be exercised in making changes to technical management techniques lest mission success be compromised.

Recommendation 1:
Transfer appropriate elements of the faster-better-cheaper management principles to the entire portfolio of space science and Earth science mission sizes and cost ranges and tailor the management approach of each project to the size, complexity, scientific value, and cost of its mission.

Science Scope and Balance

- The nature of the phenomena to be observed and the technological means of executing such observations are constrained fundamentally by the laws of physics, such that some worthwhile science objectives cannot be met by small satellites.
- The strength and appeal of faster-better-cheaper is to promote efficiency in design and timely execution—shorter time to science—of space missions in comparison with what are perceived as less efficient or more costly traditional methods.
- A mixed portfolio of mission sizes is crucial in virtually all space and Earth science disciplines in order to accomplish a variety of significant research objectives. An emphasis on medium-size missions is currently precluding comprehensive payloads on planetary missions and has tended to discourage planning for large, extensive missions.

Recommendation 2:
Ensure that science objectives—and their relative importance in a given discipline—are the primary determinants of what missions are carried out and their sizes, and ensure that mission planning responds to (1) the link between science priorities and science payload, (2) timeliness in meeting science objectives, and (3) risks associated with the mission.

Technology and Instrumentation

- Technology development is a cornerstone of first-rate Earth and space science programs. Advanced technology for instruments and spacecraft systems and its timely infusion into space research missions are essential for carrying out almost all space missions in each of the disciplines, irrespective of mission size. The fundamental goal of technology infusion is to obtain the highest performance at the lowest cost.
- The scientific program in Earth and space science missions conducted under the FBC approach has been critically dependent on instruments developed in the past. The ongoing development of new scientific instrumentation is essential for sustaining the FBC paradigm.

Recommendation 3:
Maintain a vigorous technology program for the development of advanced spacecraft hardware that will enable a portfolio of missions of varying sizes and complexities.

Recommendation 4:
Develop scientific instrumentation enabling a portfolio of mission sizes, ensuring that funding for such development efforts is augmented and appropriately balanced with space mission line budgets.

Access to Space

- The high cost of access to space remains one of the principal impediments to using the best and most natural mix of small and large spacecraft. While smaller spacecraft might appear to be the right solution for addressing many scientific questions from orbit, present launch costs make them an unfavorable solution from an overall program budgetary standpoint. Moreover, larger missions, too, are plagued by the excessive costs per unit mass for present launch vehicles.
- The national space transportation policy requiring all U.S. government payloads to be launched on vehicles manufactured in the United States prevents taking advantage of low-cost access to space on foreign launch vehicles.

Recommendation 5:
Develop more affordable launch options for gaining access to space, including—possibly—foreign launch vehicles, so that a mixed portfolio of mission sizes becomes a viable approach.

International Collaboration

- International collaboration has proven to be a reliable and cost-effective means to enhance the scientific return from missions and broaden the portfolio of space missions. Nevertheless, it is sometimes considered, within NASA, to be detrimental, perhaps because it adds complexity and can bring delays to a mission. It is also perceived to give a mission an unfair advantage and, in part, to increase NASA's financial risk.
- In the past, NASA had within its budgets an international payload line, which was an extremely useful device for funding the planning, proposal preparation, and development and integration of peer-reviewed science instruments selected to fly on foreign-led missions. This line offered the U.S. scientific community highly leveraged access to important new international missions by providing investigators with additional opportunities to fly instruments and retrieve data, especially during long hiatuses between U.S. missions in a given discipline.

Recommendation 6:
Encourage international collaboration in all sizes and classes of missions, so that international missions will be able to fill key niches in NASA's space and Earth science programs. Specifically, restore separate, peer-reviewed announcements of opportunity for enhancements to foreign-led space research missions.

1

Issues and Considerations in the Assessment of Mission Size Trade-offs in the Earth and Space Sciences

The concept of small, short-duration missions has existed virtually since the dawn of the space age. The Explorer series comprised NASA's first missions and was devoted to focused science objectives. This approach shifted in the late 1970s, when larger programs aimed at exploring the solar system (such as Viking) and programs being conceived to observe the universe in the visible spectrum (such as Hubble) and study Earth's upper atmosphere (such as the Upper Atmosphere Research Satellite (UARS)) became a focus of NASA science programs.

The space research community questioned the wholesale shift from smaller missions,[1] and there were efforts to restore a broader, more flexible approach. In 1988, the Small Explorer (SMEX) initiative harkened back to the early Explorers. It emphasized shorter programs that would allow students to participate in the development of flight instrumentation and that offered opportunities for high-priority, focused science investigations and opportunities for a principal investigator to propose and manage an entire mission.[2] Shortly thereafter, a faster-better-cheaper (FBC) approach emerged that embodied many of the practices established for SMEX and the early Explorers. Now the question is asked whether the pendulum has swung too far in the other direction, toward small, focused missions at the expense of a more mixed portfolio of small, medium, and large platforms.

This chapter outlines factors that can influence mission size and scope: (1) the laws of physics, which can impose inherent constraints for some space-based observations and measurements, (2) the scientific benefits of missions, which also entail large investments beforehand and afterwards (for example, calibration and data analysis), (3) the implementation of new management practices, (4) the role of education, (5) the need for technology development, (6) the access to space and its concomitant costs and timeliness, and (7) approaches to handling risk when implementing new management and spacecraft development practices. In addition, it reviews briefly some of the lessons learned from unsuccessful FBC missions, including the Wide-Field Infrared Explorer (WIRE), Lewis and Clark, the Tomographic Experiment Using Radiative Recombinative Ionospheric Extreme Ultraviolet and Radio Sources (TERRIERS), Mars Climate Observer, and Mars Polar Lander.

---

[1] For example, the SSB commented on the value of small missions in its report *Planetary Exploration 1968-1975*, which stressed the importance of a "series of relatively small and inexpensive spacecraft" to provide for a broad and flexible program in planetary exploration. See Space Science Board, *Planetary Exploration 1968-1975: Report of a Study by the Space Science Board, June 1968*, National Academy of Sciences–National Research Council, July 1968, p. 5.

[2] Daniel N. Baker, Gordon Chin, and Robert Pfaff, Jr., "NASA's Small Explorer Program," *Physics Today*, December 1991, pp. 44-51.

It is important to remember at the outset that the basic goal of space-based science is to answer fundamental questions about Earth and its place in the universe. It is the task of the sensing systems to gather the data required to respond to those questions. Findings are arrived at through data analysis and evaluation, are combined with other data and tools available to the investigators, and are then communicated in scientific reports. A science mission is shaped by its goals, and it is incomplete without a thorough process of data analysis.

FUNDAMENTAL SCIENCE LIMITS

Because the Earth and space sciences that utilize observations from space encompass diverse scientific disciplines, the goals for space missions are diverse as well. The scientific goals may call, for instance, for substantially different mission time horizons, orbit requirements, and size and complexity of instrumentation, and the measurements may exploit different segments of the electromagnetic spectrum. The scientific goals of a mission will also dictate measurement and instrument parameters—resolution, wavelength, repetition cycle, and area coverage, among others. The wide variety of goals and the associated instrumentation leads to a wide variety of mission complexities and spacecraft sizes. In many situations, scientific return is enhanced if the measurements from different instruments are temporally and geographically coincident, at least to within the tolerance allowed by the rate of change of the process being observed. The instruments that collect the desired data can be placed on a single large spacecraft or on separate smaller spacecraft, with the choice being guided by the scientific requirements, the cost constraints of the program, and the availability of technology for the instruments and spacecraft.

Context

Space-based instruments provide a means of collecting scientific data. These instruments are part of a larger dynamic system, shown in Figure 1.1. The observable properties[3] of a body in space or of a physical region are typically known (at least approximately). This knowledge is used during the early stages of mission design to establish data requirements that are both scientifically sound and physically feasible. One output of mission design is a set of requirements for a suite of sensors and the supporting satellite. Technological improvements, especially during the past decade or so, now enable smaller and more innovative ways to build portions of the system. However, engineering and process improvements are only part of the story. Design options for a satellite system and its sensing instruments must respect the laws of physics. In the spirit of FBC, a good mission takes advantage of the realizable design space within these limits to achieve its scientific objectives in a cost-effective manner. Mission planners must ensure that a focus on speed or cost does not obscure the subtleties and deeper issues that surround the science system paradigm.

The size and complexity of a science mission are determined to a large degree by the scope of its goals. A comprehensive set of measurements necessitates a more capable spacecraft and instrument suite than would a more limited and focused set of measurements. Examples of missions with goals both broad and narrow abound in this report.[4]

The laws of physics set fundamental limits for scientific satellite systems, thereby presenting challenges, often major ones, for any space-based science enterprise. These limits depend on the phenomena to be observed (which of course conform to the laws of physics) and the observing systems themselves (whose design must also respect these laws). The limits that dominate this design space are summarized in the following paragraphs.

---

[3]Critical properties include atmospheric conditions and the environment surrounding the spacecraft, including temperature, radiation, or a magnetic field.

[4]See also Space Studies Board, National Research Council, *The Role of Small Satellites in NASA and NOAA Earth Observation Programs*, National Academy Press, Washington, D.C., 2000.

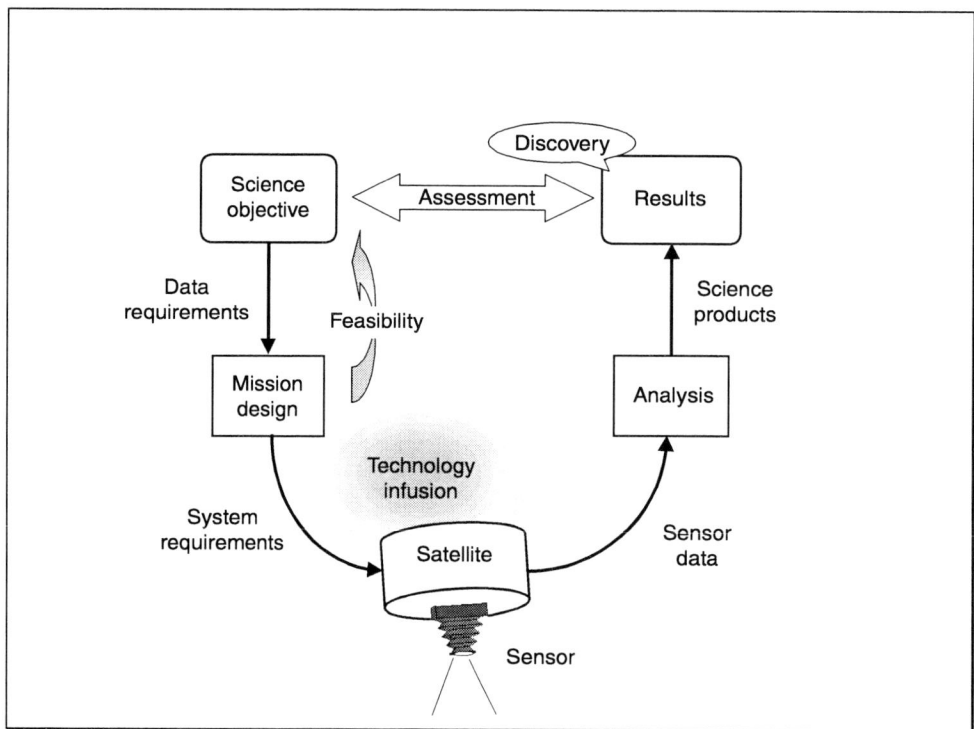

FIGURE 1.1 Conceptual model of the space science paradigm, starting from a science objective and progressing through the design of a suitable space-based measurement system to the flow of instrument data, which are transformed by thought and processing into results. Sensor and satellite implementation may take advantage of suitable technology advances. Many missions uncover new or unexpected results beyond those in the original science plan.

The Observable Phenomena

The goal(s) of a science mission establish the requirements for measurements or observations. These requirements are specific to the phenomena being measured or observed. Observing ozone depletion over Antarctica, for example, presents a very different situation (or set of physical circumstances) than measuring the gamma ray background of the universe or probing the depth of Europa's icy mantle. Each process, phenomena, or variable to be observed behaves according to its own nature. Although this statement may appear at first to be obvious, its implications are profound. There is an immense diversity among the various phenomena open to observation from space. This diversity leads to substantial differences in the size, design, and cost of the missions to observe and measure those phenomena.

The attributes of the observable phenomena that have the greatest impact on mission design and, accordingly, on mission cost can be captured by asking four questions. First, *where* is the measurement site? In general, greater distances mean higher costs. Deep space missions require more launch and spacecraft capability than low-Earth-orbit missions, all else being equal. Powerful rockets are needed to escape Earth's gravity, and fuel must be carried aboard the spacecraft to provide thrust for mid-course corrections and to assure insertion into the desired orbit at the final destination. Communications, both to and from the spacecraft, become more demanding as the distance from Earth becomes greater. Likewise, solar energy flux weakens as the distance from the Sun increases: this places greater demands on a satellite's power and thermal control subsystems.

Satellite design also must account for the environmental conditions in transit and on orbit. Extra mass and more complex subsystems are necessary to build a spacecraft that is able to withstand the Sun's heat at Mercury or to survive the charged-particle radiation surrounding Jupiter, to cite but two examples. Even in near-Earth orbit, a change in altitude of only 200 km or so may double the intensity of a satellite's radiation environment. Clusters of small spacecraft, required to separate the interlocking spatial and temporal phenomena in near-Earth space plasmas, for example, may result in a more complex, more costly, and more data-intensive mission than a single, simple spacecraft like the ones that performed the initial reconnaissance of the near-Earth space environment some 40 years ago. A space-based scientific mission thus depends critically on the location of the target site and the physical environment to be found there.

Next, *when* must measurements be made? This question raises two related aspects of data collection: timeliness and duration. Timeliness refers to the rate at which measurement opportunities occur, whereas duration refers to the length of time over which the measurements must be made. Both aspects are driven by the physics of the phenomena of interest.[5] In short, the rate of change of the observed process determines the timeliness required of the observing platform.

The duration of a measurement program is another matter.[6] Any observable phenomenon that might give rise to a cyclic signal must be observed over several complete cycles to provide enough data to characterize the underlying process.[7] Often the signal of interest, especially if it shows a long-term trend, may be deeply obscured by natural short-term variations. Attempts to gather evidence of trends in the Earth's climate are a case in point. If the mass of the polar ice sheets is increasing or shrinking, for example, this fact can be established only after accumulating a measurement time series of sufficient length that the seasonal and year-to-year weather effects can be averaged out.

Then, *what* is to be measured? The primary objective of a measurement program is to increase understanding of the observed phenomena. Often, however, the phenomena can be observed only indirectly. Most sensing instruments collect energetic particles or electromagnetic energy (such as light or radio waves). The reflective or emissive properties of each observation opportunity limit the instrumentation options.[8]

Lastly, *how* are the measurements to be transformed into scientific products? The objective of any science mission is to increase knowledge. Data collection is an essential step along the way, but it is not sufficient unto itself. For the science product to have value, generally the data need careful processing. This processing includes, at a minimum, removing artifacts and characteristics imposed on the data by the instrument and converting the engineering numbers represented by the sensor data into quantities that have physical utility. Transforming data into science products is necessarily a more complicated process if the primary data are only indirect or subtle indicators of the underlying phenomena of interest. Tighter constraints on data quantity and quality usually imply greater emphasis on the data analysis phase of a mission.

The issues raised by the last question are well illustrated by the topic of global warming. It is generally accepted that an average warming trend of only a few degrees Celsius over 100 years would induce substantial and perhaps catastrophic changes in our environment. Is there evidence for such a trend? Regardless of the answer, the question itself sets requirements for the scientific systems that would collect the necessary data. The original data must have sufficient accuracy, precision, and duration to expose the significant trends. To bolster confidence in the results, similar findings should be obtained by different means. This implies the need for several (statistically independent) observation methodologies, which in turn implies cross-calibration and a thorough understanding of the underlying physics in each case. Spatial and temporal differences between observations have to be folded in. The lessons of this example may be generalized to adapt to virtually all science questions that can be investigated

---

[5]For example, solar activity needs to be measured continuously to maximize the likelihood of observing a potentially significant but sporadic event. On the other hand, a study of the slowly enlarging deserts of sub-Saharan Africa needs to allow for seasonal variations, a requirement that could be satisfied by observations made only a few times a year.

[6]A single El Niño event can be covered by observations that span a year to 18 months, whereas it may take decades of data to establish with confidence a correlative pattern in the recurrence of El Niño events. Solar irradiance exhibits cyclic variations every 11 years or so.

[7]See SSB, *The Role of Small Satellites in NASA and NOAA Earth Observation Programs*, 2000, p. 16.

[8]For example, the Magellan mission to Venus used imaging radar techniques to map the surface of the planet. The chosen radar band—about 12 cm wavelength—was a consequence of the planet's physical properties. The constant cloud cover and dense atmosphere surrounding Venus would have prevented visible light or shorter-wavelength radar from imaging the surface, whereas longer-wavelength radar would have increased the size of the radar system and reduced the amount of detail in the final radar maps.

with satellite systems. Generating science products that respond to the original questions is an essential part of the science paradigm.

The Observing System

It would be nice if science satellites, with sufficient infusion of time and money, would shrink in cost and size and grow in capability as has digital computing hardware. Moore's law,[9] which has been borne out in practice for the past 20 years or so, states that computer chip size will shrink by a factor of two every 18 months. Unfortunately, there is no equivalent of Moore's law for the many nondigital components required by space-based sensing systems. Indeed, the lower limit on physical size of certain satellite subsystems is determined by wavelength and therefore may be dictated indirectly by the physics of the phenomena to be observed. Likewise, the limits on power, data rate, or mission duration may be dictated primarily by the observing distance. No absolute guidelines exist. However, it is unrealistic to expect that all measurement requirements could be satisfied by small instruments on small satellites having missions of short duration.

Several recurring and well-established physical principles constrain the design and performance of a satellite and its sensing systems. Five fundamental physical limits are reviewed in the following paragraphs.[10]

Kepler's laws (ca. 1620)—which are consequences of Newton's laws (1687)—elegantly describe the motion of a body subject to gravitational forces and thus describe a satellite in orbit about the Sun, a moon, or a planet. For a satellite in near-circular Earth orbit, the time required to complete one revolution is determined primarily by its altitude.[11] A satellite in Sun-synchronous Earth orbit—tuned to cross the equator always at the same local time—may be required if the science objectives for a mission require a constant angle of solar illumination, for example. Several candidate instruments may share that requirement. In such a case, there may be good reasons to mount these instruments on the same platform, resulting in a relatively large satellite.

The laws of orbital mechanics often demand a space–time trade-off: a shorter revisit interval implies less dense spatial coverage.[12] Under such constraints, it may not be possible with only one satellite to achieve the simultaneity required for fine spatial coverage and short revisit intervals. The space–time trade-off may make a constellation of satellites an appealing way to meet the science objectives of a given mission, an option that can be cost-effective only thanks to the availability of technology for smaller, more capable satellites and sensors.

Newton's laws of motion (1687) also describe the response of an individual body to its own inertia and to forces applied from internal or external sources. Both sorts of forces impact all spacecraft. Consider, for example, internally generated forces. Spacecraft that depend on solar panels for generation of electrical power may have to contend with rotations of their solar panels to keep them facing the Sun. Since every action has a corresponding reaction, this change in momentum must be compensated for.[13] The reactive force becomes larger as the moving component becomes larger or more massive, as is often true with certain remote sensing instruments. These

---

[9] R.R. Schaller, "Moore's Law: Past, Present, and Future," *IEEE Spectrum*, 34(6), 1997, pp. 53-59.

[10] The limits selected for review are summarized for the benefit of those readers who are not experienced in space system design. It is recognized that this brief introduction is incomplete and may miss many issues and subtleties such as thermodynamic or charged-particle constraints that could be central to the design of a particular space sensor or system.

[11] For example, a spacecraft in low Earth orbit (~800 km) requires about an hour and a half to go once around the world. Meanwhile, Earth rotates beneath the spacecraft, so that the ground track of the satellite gets back to the equator several thousand kilometers to the west of its previous passage. Complete coverage of Earth from such an orbit requires many days.

[12] Through choice of a satellite's altitude, its orbital period can be matched to the rotation rate of Earth, so that they rotate together. This requires a high satellite altitude, approximately 37,000 km over the equator. Seen from Earth, such a geosynchronous satellite tends to hover overhead, moving neither east nor west. This orbit is well suited for generating time series of observations over the region directly below the spacecraft, but other regions are not covered at all. Of course, the greater distance from Earth implies tougher constraints on imaging sensors and communication systems, among other trade-offs.

[13] The same principle applies to all mechanical subsystems. Thus, if there is an instrument aboard that uses an oscillating mirror to scan its field of view, as is true for many optical and infrared imaging systems, then there must be responsive reactive motions of the spacecraft on which the scanning instrument is mounted.

unwanted reactive movements have to be offset either by larger spacecraft mass or by active subsystems such as reaction wheels designed to compensate for the inertial reactive movements. Mass and complexity, of course, usually add to the cost. In general, instruments that must provide very fine angular resolution require in turn that their host spacecraft satisfy very stringent angular stability requirements.

Consider also an example of externally generated forces. All spacecraft must be launched from Earth. Newton's laws and the characteristics of available propulsion systems impose strict limits on the payload that can be lofted to Earth orbit or beyond. These limits are compactly expressed in the special form of Newton's second law of motion, known as the rocket equation. In short, flights out of Earth orbit to distant planets require substantial energy and take a long time.[14] The liftoff mass must include the extra propulsion fuel required for the spacecraft and its instrument payload to get into interplanetary transfer and final orbit insertion. Shorter transit times to planetary targets for a given payload can be achieved only if the propulsion systems are more capable, hence more massive. Larger liftoff mass requires a larger launch vehicle and more fuel. As a direct consequence, planetary missions must be more expensive than otherwise comparable Earth-observing missions.

Maxwell equations (1873) describe the behavior of electromagnetic waves as they propagate. Portions of the electromagnetic spectrum are used by all space missions. Satellite sensing and communication subsystems must be designed within the constraints of the Maxwell equations. The first and most obvious constraint is that radio waves travel at the speed of light. Even at this great speed, light travel time imposes substantial delays on all communications between Earth and satellites, especially deep space probes. Propagation time delays of 30 minutes and more are not unusual for the latter. Near-instantaneous round-trip communication is not possible. This means that planetary or deep space satellites must be designed with more control autonomy than their near-Earth counterparts. The second consequence of the Maxwell equations is that radio and light waves get weaker in proportion to the square of the distance between the radiation source and the observer.[15] This means that very distant photon sources become very faint and therefore require much larger viewing apertures.

The same physical principle impacts a satellite's electrical power subsystem. Whether intended for space physics, astronomical, or planetary missions, proportionately larger solar panels or more capable solar concentrators are required for far-ranging spacecraft if the Sun's energy is to be the main source of their onboard power. In general, solar energy is not sufficient for missions that would go to the outer planets or beyond our solar system, although developments in advanced solar arrays or concentrators might enable some outer solar system missions. For missions traveling beyond our solar system, alternative means of power generation must be found. The energy options currently available may increase the overall mission cost.

Airy diffraction (ca. 1835) enforces a lower limit on the resolution (or beam width) of any device that radiates or receives electromagnetic energy. For example, the Very Large Array (VLA) distributed-aperture radio telescope spans several kilometers of ground surface area to achieve high-resolution imagery of distant celestial radio sources. The same principle applies to spacecraft instruments, such as an optical system that is designed to image a certain level of detail on a planet's surface. The diffraction limit requires that the optical aperture diameter must be directly proportional to the satellite's distance from the surface.[16] The sensor's internal optical path length also has to grow in proportion to aperture diameter if similar performance is expected, requiring the whole instrument to be much larger.

The diffraction limit on aperture size has deeper implications as well. For any device that sends or receives energy, the size of the aperture must be proportional to the wavelength it uses. The wavelength of visible light is

---

[14]For example, the minimum-energy spacecraft path (a Hohmann transfer) from Earth to Mars takes 8.4 months and to Pluto, 30.5 years.

[15]System design takes this into account through the link budget (minimum acceptable power and gain) that describes the Earth-satellite communication subsystem. Part of this burden can be borne at the Earth-based end of the data link. This is one of the reasons that the antennas used by NASA's Deep Space Network are very large. However, the design of spacecraft communication systems also has to take this principle into account, typically requiring a fairly large dish antenna aboard interplanetary spacecraft. Proportionately greater power and larger communication antennas are required for satellites that venture farther from Earth.

[16]As an illustration, the optics on a satellite at geosynchronous altitude (~37,000 km) above Earth must have an aperture more than 40 times larger than a similar instrument on a satellite in low Earth orbit to have the same surface resolution.

very short, approximately one-hundredth the thickness of a human hair. For the much longer wavelengths used by radar systems, the diffraction limit requires apertures to be much larger than their optical counterparts.[17] To meet a given level of performance, an instrument's minimum aperture size is dictated by wavelength and distance; it cannot be reduced by technology, although distributed apertures may under some circumstances be feasible, albeit at the price of added complexity.[18]

The Nyquist frequency (ca. 1928) dictates the minimum number of digital samples per second, or sample frequency, required to transport a given amount of message detail over a communication channel. The sample frequency is also determined by how often a point on the surface must be sampled to resolve the variations in the physical processes over time. In general, more information implies more detail and more data. If those data have to be transferred rapidly, then the data rate must increase in proportion. Large amounts of data cannot be forced rapidly through low-capacity channels, as users of the Internet know only too well. Data rate requirements can drive mission costs. Sensor requirements that stipulate both large area coverage and high resolution (in time, space, and/or frequency) tend to be data-greedy. Broadly speaking, data volume and hence data rates grow in proportion to the number of instruments, to the number of channels in each instrument, to the number of resolved sample points in each channel, and to the number of digital bits required for each sample. Researchers must avoid the temptation to collect more data than are actually required to address the science issues.

The data rate required for a spacecraft communication link to the ground can be reduced if more time is allotted to transferring a given amount of onboard data. Good mission design takes advantage of this relationship.[19] Likewise, the rate required for a given signal stream can be reduced by application of data compression techniques, under suitable conditions.[20] Most deep-space missions would not be feasible without such clever mission designs. In all cases, however, the final design for the communication channel must satisfy the fundamental data rate limits.

Summary

Science objectives, seen through their attributes of where, when, what, and how, establish the data requirements. The satellite system together with its sensors has to provide those data. System design is subject to the fundamental limits imposed by the laws of physics, especially those formulated by Kepler, Newton, Maxwell, Airy, and Nyquist. The implied consequences usually emerge as lower bounds, such as limits on the energy needed to perform the mission, limits on the size of certain components central to the system, and limits on the time required to satisfy both the observation requirements and data transfers through communication links (see Figure 1.2).

There are further implications. Consider the issue of spacecraft size. Large structures such as solar panels or antennas often have to be folded if they are to fit inside the payload fairing of a launch vehicle. Larger structures,

---

[17]The main antenna for Magellan, a radar mapping mission to Venus during the period from 1989 to 1994, was 3.7 meters in diameter. Another example, NASA's Seasat imaging radar antenna (1978), was 10 meters long. If built today using the most advanced technology available, a Seasat-like antenna still would have to be 10 meters long if comparable performance were required.

[18]Enhanced performance can be enjoyed when the circumstances are favorable. For example, a thinned array may substitute for a filled array if the viewing objective is relatively sparse, such as a pair of distant astronomical objects. Such a thinned array could perhaps comprise several smaller apertures each mounted on a constellation of smaller spacecraft flying in formation. The angular resolution of such an array, however, is governed by the Airy diffraction limit, and the radiometric sensitivity of the array (proportional to the areas summed of all contributing subapertures) is governed by Maxwell equations. If the field of view is filled, as in Earth or planetary surface imaging, however, a thinned array approach usually is not acceptable.

[19]For example, images and data from the Galileo spacecraft in orbit around Jupiter were taken and recorded on board over a relatively short period and then relayed to Earth over a long downlink time, reducing the data rate required of the damaged telemetry subsystem. The Nyquist limit was satisfied, while the science content was optimized.

[20]Data compression is not always an appropriate design option. Data compression techniques require internal redundancy in the data sequence if compression is to work. An imaging radar such as Magellan's primary instrument, for example, produced data with very small internal redundancy, so that the usual data compression techniques could not be applied.

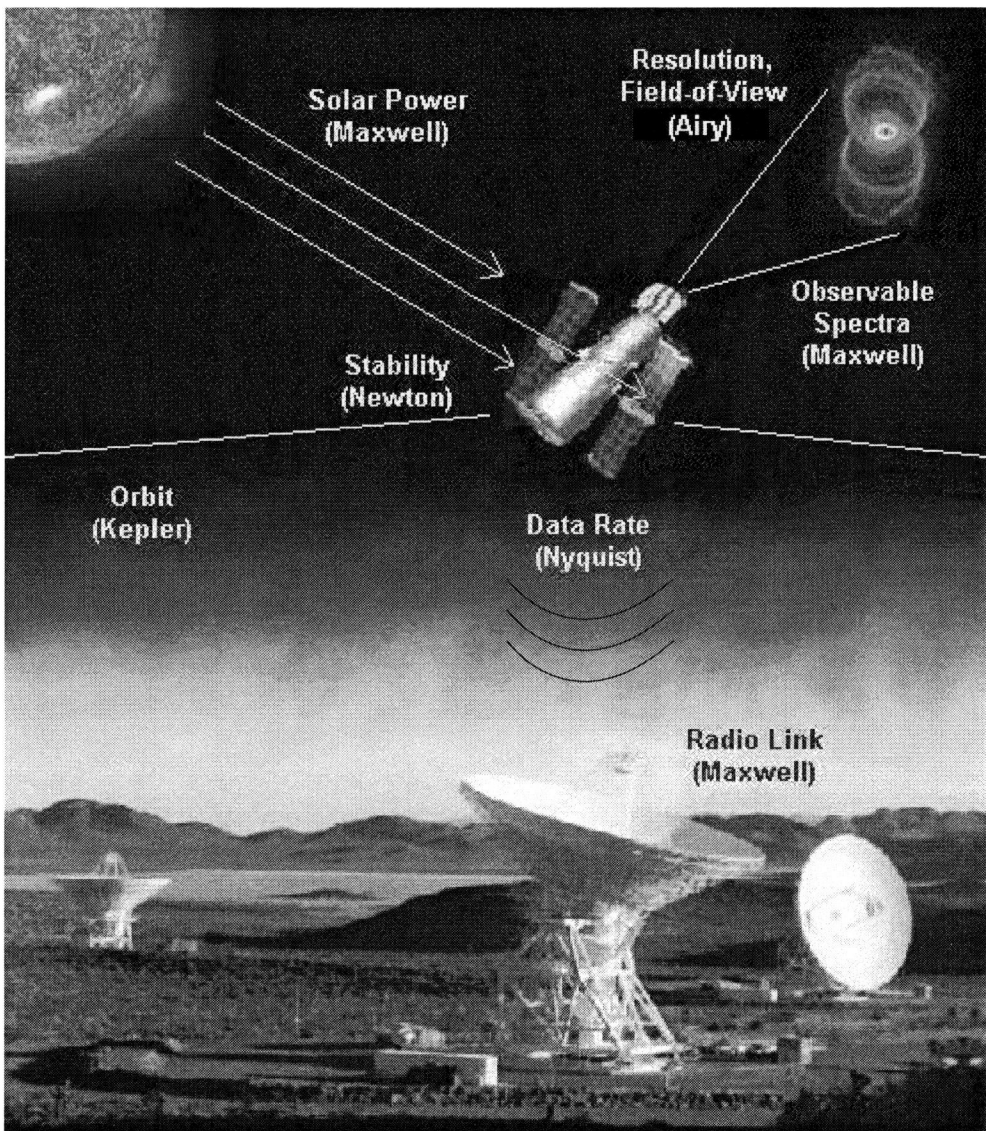

FIGURE 1.2 Depiction of the Hubble Space Telescope annotated to show critical system aspects that are subject to the principal physical limits described in the text. Hubble's size (and ancillary physical characteristics such as pointing stability and maneuverability) and relative complexity are a direct consequence of its mission, which is to provide high-quality images and spectra of very faint, distant, and finely structured features in the universe. IMAGE SOURCES: <http://www.stsci.edu>; <http://www.jpl.nasa.gov/pictures/dsn/goldstone.html>; and <http://nssdc.gsfc.nasa.gov/photo_gallery/>. CREDITS: Raghvendra Sahai and John Trauger (Jet Propulsion Laboratory), the Wide Field and Planetary Camera Science Team 2, and NASA for the Hourglass Nebula (MyCn18) shown in the upper right-hand corner.

even if folded, may dictate a larger fairing. Folded structures are meant to deploy before use, which imposes added cost and risk on any program. Much more than a relatively simple rigid assembly, a large folded structure requires special attention to its survivability in the launch environment and more extensive analysis, test, and qualification. Consider also the issue of data rate. Data collected with space-based instruments are worth little unless and until they are transferred by the data downlink subsystem to the ground. High-rate downlinks are more costly because they require more power and larger (higher-gain) antennas. High-rate links may be infeasible for planetary or deep space missions. Instrument costs can become unreasonable if the amount of data they produce is very large, as has been proven more than once.[21]

Once there is agreement on science objectives, a cost-effective system should evolve through a thoughtful mission design process. Since legitimate science objectives vary widely, well-designed science missions likewise vary in size and complexity. Absolute criteria are not appropriate when deciding on mission size. Advanced lightweight materials and fabrication techniques may reduce the mass and risk of a particular spacecraft and its systems relative to related predecessors, but there is no way to circumvent the limits imposed by basic physical principles. The role of advanced technology is to fit more technological performance and capability into a spacecraft, given the fundamental limits that govern the phenomena to be measured or observed. In the final analysis, the lower bound on the cost of a given mission is determined, directly and indirectly, by the convolution of scientific objectives with the fundamental laws of physics.

MEASURING AND ENHANCING THE SCIENTIFIC RETURN ON INVESTMENT

An important question is how small missions compare with larger, more complex missions from the point of view of scientific accomplishments; that is, what are the results for each dollar spent? This section attempts to show how the scientific return of missions of different sizes can be assessed.

Data Analysis

To recover the full scientific return on investment in space missions requires sufficient resources for calibrating (pre- and in-flight) scientific instruments, processing and archiving data in a database, analyzing and interpreting results, and publishing the findings. The issue of balance between funding for the design and development of flight hardware on the one hand and for data analysis on the other is not new and has been addressed before by the Space Studies Board and other advisory bodies.[22] It is vital that adequate funding be reserved for data calibration and analysis. In many missions, the cost margin for the project may become depleted in the early stages of the mission process, which later reduces the funds available for data analysis if the mission costs are rigidly capped. This issue is not confined to small missions; rather it is symptomatic of tighter mission budgets generally and of the fact that data analysis comes at the end of a mission, when cost overruns have consumed budget margins. There is a tendency to push data analysis into already stressed research and analysis programs. Applying faster-better-cheaper principles in ways that curtail data analysis would compromise the scientific returns from a mission.[23]

---

[21]NASA's experience in atmospheric sounders provides two pertinent examples. The Advanced Infrared Sounder (AIRS) was developed by NASA as an improved version of the advanced High-Resolution Infrared Sounder (HIRS/3) in service on NOAA polar orbiters. AIRS, to be launched only on Aqua, will achieve three times better vertical resolution and four times better spatial resolution, but its mass (140 kg) is several times that of HIRS/3. NASA's intent was that AIRS would be adopted by NOAA as an operational sensor, but NOAA was not willing to accept it. AIRS is not being continued. NASA then tried to develop the smaller and more capable Integrated Multispectral Atmospheric Sounder (IMAS), having greater measurement capacity than AIRS, comparable data rate, and smaller mass. However, the $150 million cost was prohibitive. After several years of development, IMAS was cancelled.

[22]Space Studies Board, National Research Council, *Supporting Research and Data Analysis in NASA's Science Programs: Engines for Innovation and Synthesis*, National Academy Press, Washington, D.C., 1998, p. 10; NASA Advisory Council, Space and Earth Science Advisory Committee, "The Crisis in Space and Earth Science: A Time for a New Commitment," NASA, November 1986.

[23]Space Studies Board, National Research Council, *The Role of Small Missions in Planetary and Lunar Exploration*, National Academy Press, Washington, D.C., 1995, p. 23.

Loss of data analysis funding can have a qualitatively different impact on smaller missions compared with large missions. When the RAND Corporation analyzed a set of small science missions, it found that on average (mean statistic), the resources devoted to scientific analysis represented 1.6 percent of the total mission cost.[24] The set of missions used in the RAND study does not include several small missions currently operating or under development, but the small proportion of resources devoted to science is striking.

Inadequate analysis of the data from a mission may also introduce information gaps that impair the science investigations planned for subsequent missions. Even before a spacecraft is launched, tight schedules and reduced budgets can lead to insufficient calibration of scientific instruments. Without proper calibration, any data obtained during the missions are severely degraded and limited in value.[25] In addition, the data may have to be reprocessed as corrections and/or calibrations are obtained.[26]

Smaller missions generally have more restricted goals and address a more limited range of scientific issues than do larger missions. Their data products may be relatively specialized, and in many cases means can be designed to process, analyze, and disseminate the data efficiently.[27] For larger missions, the resources available for scientific analysis may be greater. At the same time, these larger missions may also generate larger and more complex data products that place correspondingly large demands on the data analysis systems. Missions at the larger end of the mission size scale are more costly to implement, but a number of benefits may accrue—for example, the development of more powerful database and visualization tools—from having to serve a larger number of scientists.

Such tools, whether created for small or large missions, enhance the scientific impact by encouraging the wider and more timely distribution of data and the science products derived from them. "Wider distribution" means distribution beyond the core research team to other professional researchers and even other constituencies (e.g., the commercial and educational communities). Increasingly, data are analyzed or correlated with data from other missions or ground-based efforts. (One example is the Advanced Satellite for Cosmology and Astrophysics (ASCA), which developed a software system for analyzing the X-ray astronomy data that have been collected from a number of different X-ray astronomy satellites.)[28] Databases are becoming less mission-specific and better integrated into larger archives (e.g. the Planetary Data System) or, in solar and space physics, better able to meet the needs of agencies other than NASA (e.g., NOAA and the Department of Defense) for databases on the geospace environment. New insights emerge from juxtaposing separate data sets, and new ideas come from enlarging the group of people working with the data. In this way, smaller and larger missions enhance each other in terms of overall impact, justifying a mixed portfolio of mission sizes.

Similarly, the development of more powerful database tools can enhance the timeliness with which results are disseminated. This is important because even in optimal circumstances several years may be needed to recognize and follow up on patterns identified in the data sets. It is not unusual for journal articles to follow mission observations by several years owing to the need to apply improved calibrations (this need is intensified by more sophisticated databases), the need to understand subtleties in the data, and the protocols of review and publication. As an example, publications pertaining to the International Sun-Earth Explorer (ISEE) mission first peaked 4 years after the primary mission ended and then again 9 years later.[29]

---

[24]The faster-better-cheaper missions included in the data set were NEAR, Mars Pathfinder, SWAS, TRACE, MAP, Deep Space-1, Earth Observer 1, Lewis and Clark, Mars Global Surveyor, Mars '98 lander and orbiter, and Clementine. The RAND Corporation calculated that of an average mission cost of $145 million, only $2.4 million (1.6%) was allocated for scientific data analysis. See Liam Sarsfield, *The Cosmos on a Shoestring*, RAND, 1998, p. 105.

[25]Space Studies Board, National Research Council, *Lessons Learned from the Clementine Mission*, National Academy Press, Washington, D.C., 1997, p. 21; SSB, *The Role of Small Satellites in NASA and NOAA Earth Observation Programs*, 2000.

[26]For example, the National Oceanic and Atmospheric Administration (NOAA) and NASA have reprocessed ocean altimeter observations from TOPEX/Poseidon and other altimeter satellites using consistent calibration and correction algorithms.

[27]SSB, *The Role of Small Missions in Planetary and Lunar Exploration*, 1995, p. 23.

[28]Space Studies Board, National Research Council, *U.S.-European-Japanese Cooperation in Space*, National Academy Press, Washington, D.C., 1999.

[29]See SSB, *Supporting Research and Data Analysis*, 1998, p. 26.

Measuring the Scientific Return on Investment

Analysts have proposed a variety of methodologies to compare the amount of scientific data per dollar produced by smaller and larger missions. However, determining the volume of data collected from a research satellite as a way to measure the quantity and quality of the scientific output of a particular research mission is problematic because instruments differ so greatly, as do the missions themselves, and simple formulas are liable to give misleading results. Evaluating a mission's scientific productivity needs to reflect any special features or limitations of the mission acknowledged at the time the mission proposal was approved.

Figure 1.3 shows the role of scientific analysis in the end-to-end process of space-based research. As an illustration, an instrument on UARS was designed to measure the integrated solar irradiance—essentially a single number—whose absolute value and variation over the 11-year solar cycle are of extraordinary importance. It would be absurd to apply the same data-rate metric for this UARS instrument as for, say, an HST camera that takes multispectral images of distant galaxies.

There are, however, other ways to evaluate the scientific impact of a mission and to determine whether that impact is in some sense proportional to the size of the mission. Some criteria are listed here, but different weightings would be appropriate in different instances. For example, an Earth sciences mission like the Total Ozone Mapping Spectrometer (TOMS) has the potential to affect public policy, but it would be inappropriate to apply this criterion to astrophysics missions. The following are suggested as examples of criteria, not prioritized, by which the scientific impact of a mission might be judged:

- Answers fundamental questions in a scientific field, e.g., questions as outlined in NRC strategy documents;
- Leads to discoveries or other advances in knowledge that were not foreseen in the initial project proposal;
- Provides guidance or insight for the design of subsequent missions;
- Puts to effective use a guest observer (GO) program, thereby expanding the research utility of the mission. This could be measured by, for instance, the number of GOs or the configurability of the instruments and hence their attractiveness to GOs;
- Demonstrates breadth of applicability and longevity for its data products. These could be measured by, for example, the number of papers authored by scientists outside the proposing team and the number and history of file transfers from the data archives;
- Leads to a number of Ph.D. dissertations or has other substantial training impacts from its results;
- Stimulates articles in the popular press;
- Shows long-lasting legacies, e.g., textbook and encyclopedia entries; and
- Influences societal and economic issues and leads to better-informed policies.

IMPLEMENTATION

This section discusses some aspects of program implementation, including how space missions are managed, the extent of international cooperation, and the educational components of the missions.

Management

The move to apply FBC principles to NASA's Earth and space science missions has encouraged important changes in the ways NASA missions are managed. These changes have moved programs towards management practices that emphasize empowerment, leadership, and accountability.[30] In some cases, the changes have moved management into the hands of the lead scientist, making leadership by a principal investigator (PI) one mode of implementation. In the two other implementation modes currently being used by NASA—institution-led missions and PI/institution team missions—responsibility and authority are delegated to scientists and/or institutions.

---

[30]Sarsfield, *The Cosmos on a Shoestring*, 1998, pp. 26-27.

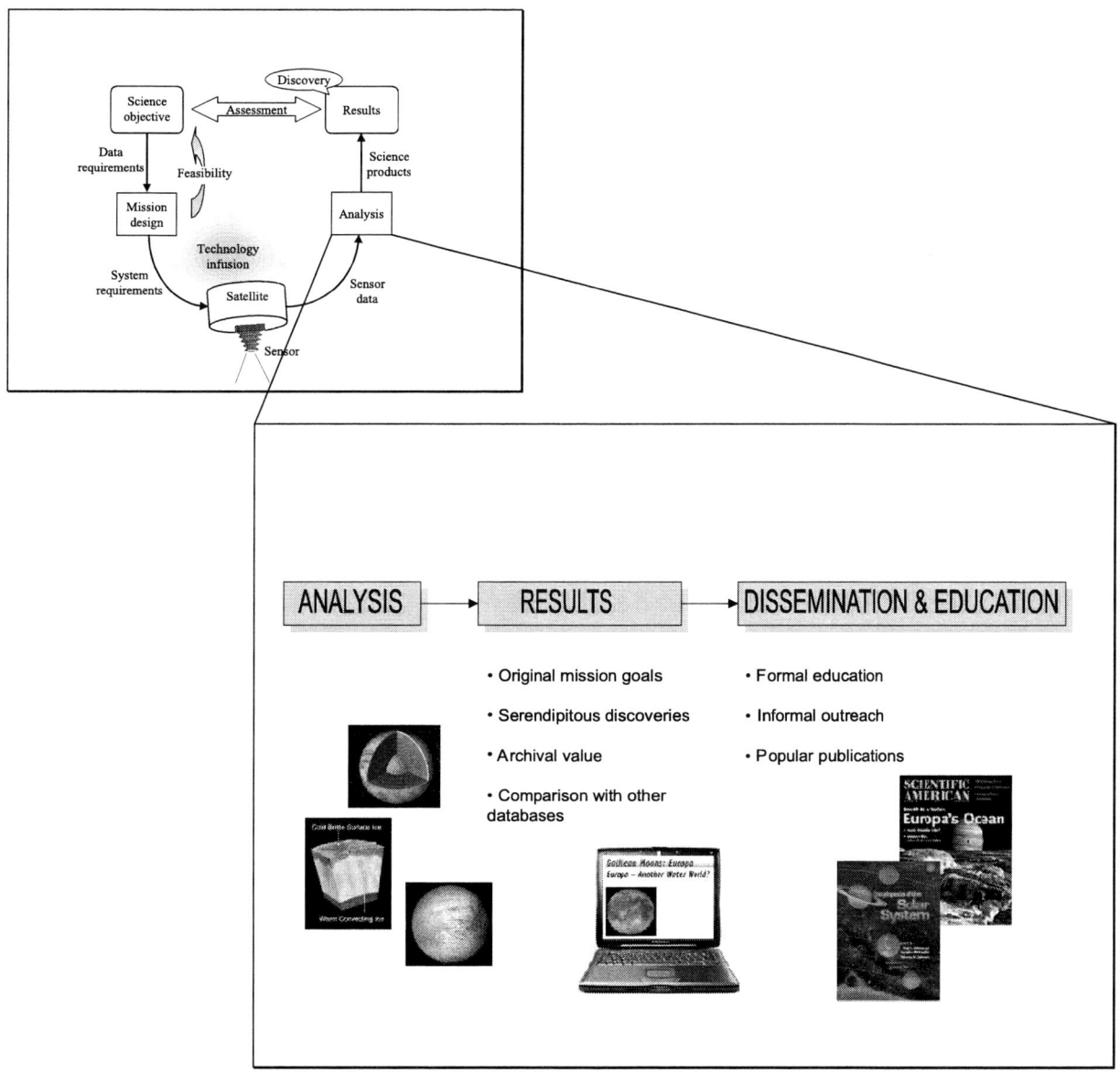

FIGURE 1.3 The raw data from a mission can have a broader impact than the originally proposed science goals. New discoveries can be made by juxtaposing data sets from different missions (e.g., observations of the same source taken at different frequencies, for example) or by taking advantage of archives to look for changes with time. The results of such analyses can be used directly in a variety of ways to promote science education, thus also enhancing the scientific impact. SOURCE: Images available at <http://galileo.jpl.nasa.gov/jupiter/jupiter.html>; cover image of Scientific American accessed from <http://www.sciam.com>, reproduced with permission from Scientific American, Inc.

Some missions, such as the Discovery, Medium Explorer (MIDEX), SMEX, and Earth System Science Pathfinder (ESSP) programs, have embraced the PI-led approach.[31] In this approach, "one principal investigator proposes an entire mission and its experiments. The idea is to have the principal investigator firmly in charge of the entire mission, with the instruments being built by the PI and his or her team of coinvestigators. The team usually includes scientists at a variety of institutions who have worked together closely in the past. The result is an efficient, highly cohesive research effort."[32] On programs such as the Student Nitric Oxide Experiment (SNOE), the PI handled mission management at the resident institution.[33] If handled properly, the PI-led approach can assure that the best FBC management practices are utilized because the close connections of the PIs with the scientific community tend to increase mission flexibility while also enhancing scientific returns.

Another approach to implementing missions of all sizes is to have a single institution such as a federal laboratory or NASA field center in charge (institution-led). This traditional approach to mission management involves a primary team consisting of a program manager to oversee the development of the mission and to manage the budget and a program scientist to ensure the scientific capability of the mission during development. In addition, a project manager and project scientist, usually from a NASA center, deal with all project development issues (such as schedules, costs, and scientific trade-offs) in close liaison with headquarters. The institution-led approach can provide continuity of management and a corporate memory from mission to mission. Other benefits include access to the resources and infrastructure required for handling larger, more complex missions, the ability to bring in expertise to resolve a technical crisis, and efficiency in employing technical staff and managers. Missions such as Mars Pathfinder and Magellan (both JPL missions) and the Earth Observing System (Goddard Space Flight Center) exemplify the institution-led mode.

The third implementation mode, the PI/institution team approach, involves a PI-led mission that partners with an institution such as a NASA center. In this case, the PI may choose not to lead the mission independently, may require management support, or may wish to incorporate contributions such as management expertise, infrastructure, or facilities for mission development from partnering institutions. Stardust, a mission collecting samples of interstellar dust, is an example of the PI/institution-team approach. Stardust involves a PI at the University of Washington who oversees the mission; JPL, which provides the project management team; and Lockheed Martin Astronautics, which built the spacecraft.[34]

FBC principles and streamlined management practices can and should be applied in all three implementation modes, regardless of the size of a mission, to achieve shorter development periods, cost-effectiveness, accountability, and empowered decision-making. For example, FBC principles are being applied in developing the Space Infrared Telescope Facility (SIRTF) mission, which offers the chance of monitoring the application of such principles to larger missions. In addition, pursuing efficient contractor practices, low-overhead management techniques, concurrent engineering, integrated product development teams, and fewer formal reviews as well as coordinating with rather than overseeing the contractor, can enhance management effectiveness.[35]

The foregoing suggests that FBC principles applied wisely are beneficial for all management approaches and implementation modes. In addition, the diversity of modes for executing missions, coupled with the increasing frequency of spaceflights under the FBC approach, is critical for encouraging industry, universities, NASA centers, and government laboratories to participate actively in the space research program. Each player offers assets (e.g. innovation, infrastructure, intellectual capital, engineering capabilities, management expertise, development experience) that when taken together with those of the other players, strengthen the health and development of the space research program. The modes for managing missions (the PI mode, the institution-led mode, and

---

[31] Several Space Studies Board reports have pointed to the attributes of management approaches adopted for smaller, shorter-duration missions. See Space Studies Board, National Research Council, *Scientific Assessment of NASA's SMEX-MIDEX Space Physics Mission Selections*, National Academy Press, Washington, D.C., 1997. See also SSB, *Clementine*, 1997; SSB, *Small Missions*, 1995; SSB, *Small Satellites in NASA and NOAA Earth Observation*, 2000.

[32] Baker et al., "NASA's Small Explorer Program," pp. 44-51.

[33] See <http://lasp.colorado.edu/snoe> and <http://cass.jsc.nasa.gov/stedi/overview.html>.

[34] See <http://stardust.jpl.nasa.gov/mission/msnover.html>.

[35] SSB, *Small Satellites in NASA and NOAA Earth Observation*, 2000.

the PI/institution team mode) capitalize on these various strengths. For instance, the PI mode focuses on the university and science community, while the institution mode focuses on government laboratories or NASA centers and the resources and skills they offer. Approaches and mixes of players for implementing missions are changing and may continue to incorporate traditional and FBC management practices.

There are, however, limitations and caveats for all the management modes. As an example, experience from the Lewis and Mars Climate Observer (MCO) missions (see section "Problems with Past Missions") has shown that streamlined management methods must receive sufficient oversight and attention to avoid pitfalls.[36] As one moves up the scale of mission size, the engineering "tail" can begin to "wag the scientific dog," and sometimes costly attention to engineering details is required to contain risks. The Lewis and MCO failure investigations showed that the cost and schedule constraints associated with the FBC approach created stresses that resulted in poor management decisions. A management team of appropriate size and experience is needed to reduce the probability of mission failure caused by such problems as too few technical checks, poor communications, or inappropriate workload allocation.[37] Other concerns are the effectiveness of the PI mode for medium-scale missions and the need for instrument hardware to support PI-led missions. The PI mode has not yet been evaluated for MIDEX-class missions[38] and has so far been applied at a time when PIs have had access to the scientific instruments they need to conduct smaller, shorter-duration missions. Much of the cost of developing these instruments has already been expended and is not included in the PI's mission budget. The effectiveness of the PI mode may change when new instruments must be conceived, developed, and tested.[39]

International Collaboration and FBC Principles

The mission portfolios of NASA's Earth science and space science programs can benefit substantially when opportunities for international collaboration are taken advantage of. This section discusses the selection and planning processes for international cooperative missions, the risks in conducting such missions, and opportunities to facilitate them.

The planning processes for missions influence the ease with which countries and agencies can collaborate on space research. Other factors, such as the process for selecting missions, the ability to launch them, and the availability of resources for collaborating in foreign-led missions, are also important in conducting international collaborations. The trend in the United States towards using smaller, shorter-duration missions may be impairing the ability of international partners to participate in U.S.-led missions because of the shortened planning and selection periods.

In the case of larger missions, the planning process is likely to be relatively long and to follow a well-defined procedure. Planning can occur at the agency level and can be timed to fit the planning and funding cycles of potential international partners, although it does not always proceed in this fashion.[40] Furthermore, collaborations even on large missions can be simpler if only two countries rather than many are involved, such as on the German-U.S.-U.K. Roentgen Satellite (ROSAT) mission or on the U.S.-French Ocean Topography Experiment (TOPEX)/Poseidon mission (see Box 1.1).

Some smaller NASA missions can incorporate collaboration with a single foreign country more simply than large missions, which typically involve cooperation between large agencies such as NASA and the European

---

[36]The PI/institutional approach works when the PI has a sufficiently well-defined payload to select an appropriate bus (and hence teammate) at the time of the procurement. Most proposals submitted in response to recent PI-mode announcements of opportunity (AOs) involve payloads that have been under development for some time and are relatively mature. For example, the fourth Discovery mission—Stardust—carries aerogel capture cells proven on numerous Get Away Special Sample Return Experiments and a camera that uses spare parts from the Voyager and Galileo missions. See SSB, *Small Satellites in NASA and NOAA Earth Observation*, 2000.

[37]See Mars Climate Orbiter Mishap Investigation Board, *Phase I Report*, November 10, 1999; Robert Lee Hotz, "String of Missteps on Doomed Orbiter," *Los Angeles Times*, November 11, 1999; Sarsfield, *Cosmos*, 1998, pp. 34-35.

[38]SSB, *SMEX-MIDEX*, 1997, p. 15.

[39]SSB, *Small Satellites in NASA and NOAA Earth Observation*, 2000.

[40]Space Studies Board, National Research Council, and European Science Foundation, *U.S.-European Collaboration in Space Science*, National Academy Press, Washington, D.C., 1998, pp. 101-111.

BOX 1.1
Two International Collaborations

Ocean Topography Experiment

A bilateral mission between France and the United States, the Ocean Topography Experiment (TOPEX/Poseidon) is dedicated to observing Earth's oceans and to providing global sea-level measurements of unprecedented accuracy.

TOPEX/Poseidon, launched on August 10, 1992, weighed some 2,500 kg. NASA's share of the cost for the large-scale, bilateral mission is approximately $450 million (life-cycle costs). NASA contributed the satellite bus, five scientific instruments, and the development of the data system, and the French Space Agency contributed the launch and the associated ground processing infrastructure.

TOPEX/Poseidon is considered both a collaborative and a scientific success. Its data are used to help scientists determine global ocean circulation and understand how the oceans interact with the atmosphere. This interaction is essential to improving our understanding of global climate and other aspects of global environmental variability and change.[1] Other uses of the ocean altimetry data include monitoring large-scale ocean circulation and investigating subtle signals from tidal energy as it moves around the globe. Moreover, TOPEX/Poseidon data have been used to study the sea-level changes associated with El Niño events: the data help researchers to analyze the interaction between swirling eddies and the ocean currents that give rise to El Niño phenomena and to better understand how heat storage in the ocean changes from season to season.[2] TOPEX/Poseidon is continuing to provide ocean altimetry data and will be succeeded by a follow-up U.S.-French mission, Jason-1, to be launched in 2001.

Advanced Satellite for Cosmology and Astrophysics

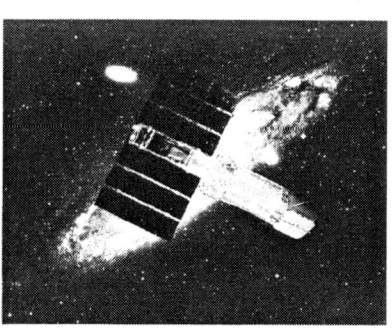

How are stellar magnetic fields generated? What physical processes determine the mass functions of single and binary stars, from their formation to their demise as compact stellar remnants such as white dwarfs, neutron stars, and black holes? What elements are ejected in novae and supernovae? The Advanced Satellite for Cosmology and Astrophysics (ASCA) is helping scientists better understand high-priority questions on stellar evolution and the origin of elements identified by astronomers and astrophysicists.

Led by the Institute of Space and Astronautical Science of Japan, with contributions from the United States, ASCA was launched on February 20, 1993, to perform spectroscopic imaging of cosmic, high-energy phenomena in the 0.5 to 10 keV band. Japan provided the spacecraft, launch, management, and operations; the United States cooperated on scientific payloads. ASCA builds on observations made by a much smaller mission, the EUVE, and has revealed dense coronal structures and reconnection and flaring of magnetic fields in young stars.

In addition, ASCA, the first true imaging spectrometer in the X-ray band, directly imaged supernovae remnants of individual atomic transitions of recently expelled elements. These data, combined with higher-resolution imaging being conducted by the Satellite per Astronomia in Raggi X (SAX), are providing, for the first time, an opportunity to determine the composition of material ejected from the cores of massive stars.[3]

---

[1] Space Studies Board, National Research Council, and European Science Foundation, *U.S.-European Collaboration in Space Science*, National Academy Press, Washington, D.C., 1998, pp. 84-86.
[2] See <http://topex-www.jpl.nasa.gov/>.
[3] Space Studies Board, National Research Council, *A New Science Strategy for Space Astronomy and Astrophysics*, National Academy Press, Washington, D.C., 1997, pp. 23 and 28.

Space Agency (ESA). Smaller missions are generally chosen through open announcements of opportunity (AOs)[41] in which numerous missions are considered but only a few are actually selected for flight. In this situation, international arrangements are informal, and because the response time to an AO is usually short (e.g., less than 4 months), a preexisting relationship with one's collaborators is generally required. The proposing PI arranges with scientists in other countries for an instrument or detector contribution, for example, and the scientists in those countries, in turn, approach a national funding agency for support if their team is selected.[42]

The AO process for small missions may have an impact on international cooperation for some missions owing to the disciplinary breadth of the announcements, which results in a large number of proposals being submitted. The odds of any one proposed mission being selected can be discouragingly small,[43] which may make discussions between foreign scientists and their national funding agencies problematic. Furthermore, whether or not international collaboration is encouraged or discouraged during mission selection seems to vary across NASA's science program offices. In some cases having a foreign partner is considered as a way to enhance the capabilities of the mission at little cost to NASA (e.g., TOPEX/Poseidon for Earth sciences or ASCA for astrophysics) (see Box 1.1). In other cases, international collaboration might be considered detrimental by some at NASA because it could bring added complexity and potential delays to a mission, because it is perceived to give a mission an unfair advantage, or because it increases NASA's financial risk.

Despite the potential benefits of cost-sharing and increased scientific opportunities, contributions from foreign partners can introduce risk into collaborative ventures. A partner could, for example, lose funding or fail to meet mission schedules, or its contributed hardware or software might not perform according to requirements or expectations. These potential risks can be incorporated into management plans and be handled with appropriate contingency plans.

International participation in NASA missions is only one side of this issue. There are also opportunities for U.S. scientists to participate in missions mounted by other nations, usually by providing spacecraft subsystems, detectors, or instruments. In general, such participation involves a relatively small funding requirement from the United States that is offered as a mission of opportunity (MoO) line in Explorer AOs. MoOs apply to satellites mounted by industry, the military, or foreign countries. While the MoO line provides opportunities for cooperation, it falls short of enabling the scientific community to integrate cooperative activities into the strategic planning process.

The international payload line that NASA used in the past offered a more systematic approach to encouraging international cooperation on foreign-led missions and to integrating such activities into NASA's planning process. NASA issued AOs to the science community for European- and Japanese-led missions (at present, researchers must locate opportunities on foreign-led missions on their own). The international payload line[44] facilitated science planning, proposal preparation, and the development and integration of instruments for the scientists who were selected for such opportunities in return for scientific data from the mission. The cooperative venture could then be integrated into NASA's strategic planning process as a way to meet certain scientific objectives in a field.

International cooperation at a variety of levels is important. A mix of mission sizes can be achieved not only by the missions in NASA's portfolio but also by U.S. participation in international missions. The ESA astronomy mission Planck is a good example. Its cost to NASA is small but its science is comparable to that of a medium or large mission.

Education

NASA science missions provide opportunities for a wide variety of educational activities. These activities take forms that range from formal education at schools and colleges to informal life-long learning, through the

---

[41]SSB and ESF, *U.S.-European Collaboration*, 1998, p. 24.

[42]While this arrangement is more likely to be possible in Europe than in Japan, where funding for PIs and guest investigators is generally at the agency level, "the specific focus and time constraints presented in NASA's Announcements of Opportunity for 'faster, cheaper, better' missions make it difficult for Europeans to respond and participate." SSB and ESF, *U.S.-European Collaboration*, 1998, pp. 23-28.

[43]SSB, *Supporting Research and Data Analysis*, 1998, pp. 54-55 and 108.

[44]SSB, *Supporting Research and Data Analysis*, 1998, pp. 61, 65-66.

mass media, science museums, and the Web. Educators with strong scientific backgrounds coordinate activities on both large and small missions with local schoolteachers. The educators bring the space missions into classrooms, public libraries, and museums using electronic media and audiovisual materials and organize school visits to witness the development and construction of a spacecraft and the operation of a mission center. At the college level, undergraduate and graduate students can become more directly involved in missions, working with scientists and engineers on the design, construction, and operation of spacecraft as well as on analyzing the scientific data returned.

Larger missions such as Voyager, Hubble, Galileo, Cassini, and the Earth Observing System (EOS) mission Terra have generated considerable public interest and inspired extensive educational activities. The presence in a community of an institution such as the Space Telescope Science Institute, the Jet Propulsion Laboratory, or a NASA center allows for substantial, formal links to the community's schools and educational networks. While the longer duration of large program missions might preclude a student from experiencing the whole sequence, it could offer more opportunity for him or her to become involved in data analysis, as long as such analysis is fully funded.

Previous Space Studies Board reports raised expectations that the FBC approach would enhance the educational role of NASA science missions because small missions need not be done exclusively at large contractors or government facilities: "Small missions provide a variety of opportunities for education at K-12 levels. The involvement of universities in small missions is also an excellent chance to excite and inspire students in various disciplines at both undergraduate and graduate levels and to provide technical, scientific, and managerial experiences that might be extremely valuable for a wide variety of careers. The most desirable missions for student participation are those that are completed in at most a few years, so that students can be a part of the entire mission—not just analyze data obtained a decade earlier."[45,46] Examples of successful student involvement in small missions (see Box 1.2) are provided by the Solar Mesosphere Explorer and SNOE at the University of Colorado and the Extreme Ultraviolet Explorer (EUVE) at the University of California at Berkeley. On an even smaller scale, the NRC report *Supporting Research and Data Analysis in NASA's Science Programs* cites the special educational opportunities that are provided by suborbital balloon or rocket flights.[47]

The committee is not aware of any assessments to date of the relative impact of educational activities associated with missions of different sizes. Furthermore, it has not seen any evaluations of whether the stipulation in an AO of an educational component for smaller, shorter-duration missions (generally at the level of one to a few percent of the total mission budget) has enhanced student involvement or educational value. There are examples of excellent education programs associated with both large (e.g. Hubble) and small (e.g. SNOE) missions, but generalizations about the overall educational benefit of large vs. small missions in aggregate are anecdotal. Comparisons are complicated by the fact that different missions have concentrated on different areas of education. Some have specialized in K-12 education (e.g., Cassini), teacher training (e.g., Kuiper Airborne Observatory), college students (e.g., Cooperative Astrophysics and Technology Satellite (CATSAT)), science museums (e.g., International Solar-Terrestrial Physics program), or the mass media (e.g., Hubble). In some cases, efforts labeled "education" are little more than public relations.

In other cases, mission management may wish to involve more students in smaller, shorter-duration missions but may not be able to do so because of the start-up costs or concerns about perceived added risk. In the drive to cut mission costs, the later parts of the mission (data analysis and synthesis) are often severely curtailed. If such

---

[45]SSB, *Small Missions*, 1995.

[46]There are few satellite instrumentation development facilities in academia for Earth science. The influences of FBC and commercial interests are naturally leading to the "commoditization" of Earth remote sensing. Such a development style requires enormous capital investment in facilities and in the training and education of people to maintain state-of-the-art capabilities, which are essential to reduce costs. Many universities have been reluctant to make these investments, and the federal government, including NASA, has been hard-pressed to do so. The emphasis for Earth science may shift to adding value to commercially developed sensors (through onboard intelligence and post-processing, for instance).

[47]SSB, *Supporting Research and Data Analysis*, 1998, pp. 31-33.

BOX 1.2
Student Participation in Space Science Missions

Student Explorer Demonstration Initiative

The Student Explorer Demonstration Initiative (STEDI) is designed to involve students intensively in the design, building, and operation of small spacecraft. At a cost of less than $10 million each, STEDI missions are the smallest in the Explorer program and the next step up in mission capability from a sounding rocket. They were nominally to be launched on Pegasus vehicles.[1] Funded by NASA and managed by the Universities Space Research Association (USRA), the program aims to demonstrate the potential for university-led teams to successfully carry out high-quality space science and technology missions at a relatively low cost on a time scale of 2 years from go-ahead to ready-for-launch.[2]

The three missions selected were (1) the University of Colorado's SNOE, (2) Boston University's TERRIERS, and (3) the University of New Hampshire's CATSAT:[3]

• *SNOE*—The University of Colorado's Laboratory for Atmospheric and Space Physics teamed with Ball Aerospace on this project. The Student Nitric Oxide Experiment (SNOE) was designed to measure nitric oxide density in the lower thermosphere (90 to 200 km altitude) and to analyze the solar and magnetospheric influences that create it and cause its abundance to vary dramatically. SNOE was designed to operate on orbit for one year. It continues to function perfectly since its successful launch on February 26, 1998.

• *TERRIERS*—Boston University's Tomographic Experiment using Radiative Recombinative Ionospheric EUV and Radio Sources (TERRIERS) attempted to utilize space- and ground-based instruments to make daily global upper atmospheric measurements. Using tomography to measure ultraviolet light emissions, TERRIERS was to survey the upper atmosphere and utilize the technique to study ionospheric/thermospheric processes. Unfortunately, TERRIERS failed (see "Problems with Past Missions"); however, it nonetheless contributed significantly to education.

• *CATSAT*—The Cooperative Astrophysics and Technology Satellite (CATSAT) of the University of New Hampshire's Institute for the Study of Earth, Oceans, and Space is designed to measure distance and polarization of gamma-ray bursts in an attempt to determine the origin of the mysterious gamma-ray burst phenomenon.

As the Space Studies Board noted in its report *Scientific Assessment of NASA's SMEX-MIDEX Space Physics Mission Selections*, "STEDI appears successful in providing hands-on educational opportunities for both graduate and undergraduate students in engineering and software development which makes students highly marketable after graduation."[4] Indeed, SNOE involved over 150 students, including some 80 paid positions for undergraduates and graduates and 70 other positions, including 10 high school students. Student responsibilities ranged from project responsibilities (mechanical and electrical designs, software development, provision of ground support equipment), mission operations (real-time command operations, orbital tracking), and data analysis (database management, data reduction, model development).[5]

Space Telescope Science Institute

The Space Telescope Science Institute (STScI) is responsible for the scientific operation of the Hubble Space Telescope (HST), including the selection and support of telescope users, the scheduling of telescope observations, the archiving of data, and planning for new telescope instruments.

continued

> **BOX 1.2 Continued**
>
>
>
> The educational and outreach charge for STScI is to conduct a national program that brings the scientific results and technology of HST to teachers, students, and the general public. Additional educational activities include development of a suite of classroom tools for use in hands-on science and mathematics education and pilot programs to encourage the partnership of research astronomers and technologists with teachers in grades K-12.[6] In addition, STScI coordinates programs for approximately 50 undergraduate, graduate, and postdoctoral students that provide financial support and/or research mentoring each year.[7]
>
> The Hubble Space Telescope was launched on April 24, 1990. It is capable of imaging objects up to 14 billion light-years away and is the largest on-orbit observatory ever built. Hubble was a joint venture between the European Space Agency and NASA. The observatory has "revolutionized observation astronomy by providing crisp images of objects ranging from protoplanetary disks and exploding stars to images of the most distant galaxies ever observed."[8]
>
> ---
>
> [1]Space Studies Board, Board on Atmospheric Science and Climate, National Research Council, *An Assessment of the Solar and Space Physics Aspects of NASA's Space Science Enterprise Strategic Plan,* National Academy Press, Washington, D.C., 1997, p. 8.
> [2]"Student Explorer Demonstration Initiative," available at <http://cass.jsc.nasa.gov/stedi/overview.html>, September 21, 1999; "USRA and the STEDI Program," available at <http://lasp.colorado.edu/snoe/descr/STEDI.html>, September 21, 1999.
> [3]"USRA and the STEDI Program," available at <http://lasp.colorado.edu/snoe/descr/STEDI.html>, September 21, 1999, or <http://cass.jsc.nasa.gov/stedi/overview.html>, September 21, 1999.
> [4]SSB, *SMEX-MIDEX*, 1997, p. 13.
> [5]Stan Soloman, University of Colorado, Laboratory for Atmospheric and Space Physics, September 1999, telephone interview.
> [6]"Amazing Space," available at <http://amazing-space.stsci.edu/educators/internflyer.html>, September 29, 1999.
> [7]Mario Livio, Space Telescope Science Institute, December 2, 1999, telephone interview.
> [8]SSB, *Supporting Research and Data Analysis,* 1998, p. 18.

cuts occur, they affect the translation of raw scientific data into user-friendly databases, which are accessed widely via the Web by scientists, students, and the general public. Loss of data analysis resources not only reduces the scientific return from the mission but also eliminates the area (data analysis) in which graduate students are most extensively involved.

TECHNOLOGY

Technologies Enhance Instrument and Bus Capabilities

Scientific research using spaceborne instrumentation has been characterized by a continuing, almost explosive improvement in the capabilities of the instrumentation as well as in the capabilities of the spacecraft bus carrying the instruments. The performance of today's instrumentation and spacecraft greatly surpasses the performance of those of only a decade ago. This growth in performance is based on a continuing investment in technology development: to prepare for the future it is essential to push even further the limits of current technology.

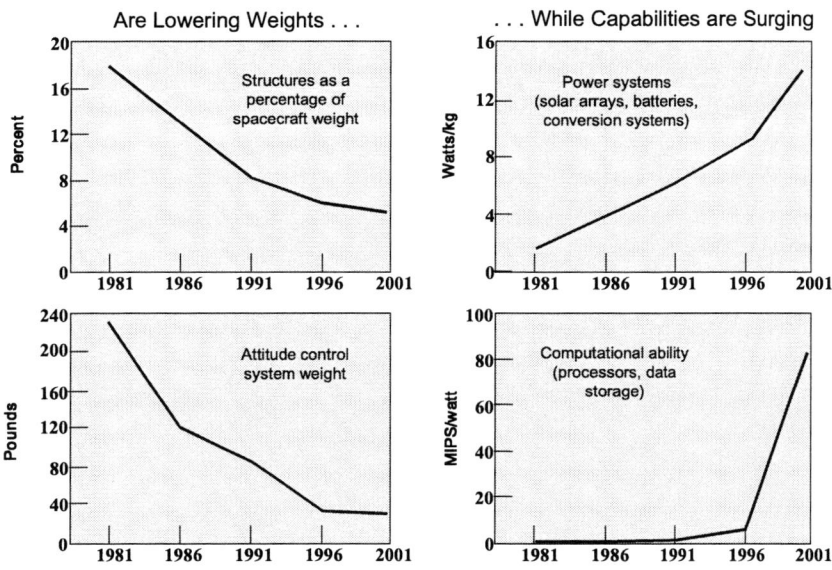

FIGURE 1.4 Advances in spacecraft capability, 1981 to 2001. MIPS, million instructions per second. SOURCE: Pedro Rustan, *Aviation Week and Space Technology*, January 25, 1999. Reprinted with permission from *Aviation Week and Space Technology*.

There is a synergism between the development of spacecraft bus technologies and technologies that improve the scientific instrumentation that enables novel space missions and greatly improves the scientific return. Advances in spacecraft technologies, such as energy-efficient electronics, more powerful solar panels, more accurate pointing capabilities, higher rates of data transmission, and onboard autonomy, increase the capabilities of instrumentation. In turn, improved instrumentation that can take advantage of the characteristics of advanced spacecraft enhances the scientific productivity of space missions.

Figure 1.4 illustrates the growth in spacecraft technology capability in the last two decades. A similar rate of progress is occurring in many of the technologies associated with scientific instrumentation, such as more capable processors and more sensitive receivers of electromagnetic radiation. In the future we will probably see the use of micromechanical systems aboard spacecraft, the deployment of large, inflatable structures and antennas, and the ability to fly many small spacecraft in formation. These technologies will lead to new research opportunities and enable novel space missions. The time scales of continuing technology improvement in some instances are relatively short, a few years or less. Space mission development and execution time scales should be comparable to the time scale of technology improvement, since space missions are expected to fly the best technology available.

Two Approaches to Infusing New Technology

The process of infusing new technology into Earth and space science missions has changed dramatically over the course of the space program. In the rush to establish the U.S. space program in the late 1950s, government and industry were willing to use any technology that was available in these early spacecraft, because technology developed specifically for space applications was scarce. Considerations of risk and cost were secondary, and speed of execution was essential. This venturesome approach to space exploration was responsible for the very successful U.S. civil space program of the 1960s and the early 1970s.

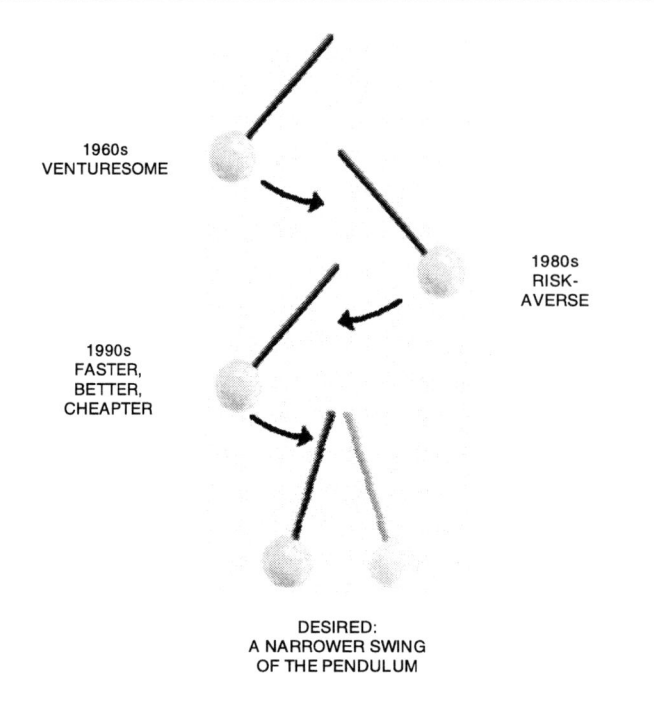

FIGURE 1.5 The pendulum of space and Earth science technology application.

As space technology evolved and space systems became more complex, program managers became more cautious about the flight qualification status of the technology being used in spacecraft. Standards for qualification tests emerged as a result of mission failures. Testing became extensive—and expensive—and the community started to demand flight "heritage" for all technology systems selected for flights. The trend to bigger, more capable satellites using flight heritage technologies became the norm in the 1980s, when the country was still enjoying an increase in the government space budget, driven by the Cold War. The U.S. budget for NASA and the Department of Defense space activities in the late 1980s was considerably larger than a decade before, and new technologies were developed for a wide variety of space applications.[48]

The confluence of long technology development times, increasing complexity, and growing costs of space systems led to an interest in cutting the time elapsed between the ground qualification of a new technology and its subsequent integration into space missions. Opportunities also arose to take advantage of the increased performance capabilities and lower masses inherent in the new technologies and to develop very capable, relatively small spacecraft. The contributions of the Clementine, the Solar Anomalous and Magnetospheric Particle Explorer (SAMPEX), the Submillimeter Wave Astronomy Satellite (SWAS), Mars Pathfinder, and Mars Global Surveyor missions, all of which used these new technologies, caused the technology pendulum to swing back in the other direction (see Figure 1.5), toward rapid mission execution, which had been the philosophy in the early days of the space program. Today, in light of the Lewis, MCO, and Mars Polar Lander failures, described in the section "Problems with Past Missions," the committee believes the pendulum has swung too far in that direction, toward practices that rate speed and cost over risk and sound management.

---

[48]NASA, *Aeronautics and Space Report of the President: Fiscal Year 1998 Activities*, 1999, p. 91.

ACCESS TO SPACE

Both U.S. (e.g., Pegasus, Athena, Titan-4) and foreign (e.g., Ariane) launch vehicles have experienced significant problems during the last several years. New launch systems have always been plagued by low reliability levels when they are first used. Therefore, any assessment of launch vehicle risk must take system maturity into account. The Space Studies Board recently noted as follows in its report *The Role of Small Satellites in NASA and NOAA Earth Observation Programs*: "... early experience with the new small launch vehicles has included a number of failures, and the present paucity of reliable options is of great concern. This is likely due in part to the relative newness of these systems and a desire to minimize development costs for these commercial ventures. Continued development should overcome the difficulties and yield a suitable balance between cost and reliability." In addition, the report noted that "present launch vehicle performance capabilities do not effectively span the range of potential payloads. For example, there is a significant gap in capability between the Pegasus-Athena-Taurus launch vehicles and Delta II."[49]

The issue of access to space is a matter of continuing concern in planning an effective, balanced, and cost-effective program for space and Earth science. Whether NASA provides access by an appropriately priced, expendable launch vehicle dedicated to a single science mission or by a secondary payload opportunity, the agency must deal effectively with launch issues. The high cost, lack of options in vehicle size and capability, and the sometimes poor reliability of launch vehicles are the principal impediments to using the optimum mix of spacecraft sizes. For example, while smaller spacecraft might seem to be the right technical approach to answering many scientific questions, present launch vehicle costs and availability discourage this approach from an overall program resource standpoint.

Even for large missions (e.g., major planetary spacecraft), the cost of launch vehicles severely constrains the science program. With launch vehicles costing many tens to hundreds of millions of dollars, exciting mission ideas face daunting obstacles. It is as true now as ever that the costs of access to space, however defined, must come down to permit an effective science program. This problem is not unique to space and Earth science; it applies as well to commercial efforts.

Finally, the ad hoc committee notes that national space transportation policy requires all U.S. government payloads to be launched on vehicles manufactured in the United States.[50] Moreover, U.S. commercial spacecraft requirements are driving the market towards larger launch vehicles, and there is little industry interest in providing opportunities for science missions to fly as secondary payloads on large commercial launchers. The ban on foreign launchers restricts the access to space of U.S. scientific payloads and prevents the science community from taking advantage of economical access offered by foreign launchers. These two factors combine to unnecessarily handicap the FBC approach.

RISK

Traditional approaches to developing missions reduced risk by using flight-qualified hardware, internally redundant subsystems, and redundant spacecraft (e.g., Viking 1 and 2 and Voyager 1 and 2). The budgets for missions conducted under the FBC approach do not afford such redundancy and therefore introduce higher risks. In addition, FBC principles, including streamlined management and the use of advanced technologies (discussed in the subsections on management and technology), make risk a bigger challenge in developing science missions. The four most important risks to consider are the following:

1. Risk in technology selection—Missions using state-of-the-art technologies in new applications tend to have higher failure rates because the technologies have not yet been flight-qualified. Regardless of the extensive simulation and ground testing performed to qualify new technologies, the space environment is not benign, and it

---

[49]SSB, *Small Satellites in NASA and NOAA Earth Observation*, 2000, p. 3.
[50]National Space Transportation Policy, NSTC-4, August 5, 1994.

is impossible to simulate all space conditions (e.g., thermal, vacuum, radiation, and electromagnetic interference) and launch conditions (e.g., shock and vibration) in tests alone. Advanced technology that lacks a flight heritage will carry an inherently higher risk. As an example, the decreased size of modern microelectronics may make these technologies more susceptible to some types of space radiation.[51]

2. Risk in payload configuration—When spacecraft size and complexity grow in response to more ambitious science goals, additional risks are incurred by having to integrate complex, multi-instrument payloads. Program managers have tended to underestimate the time required for integrating complex payloads and the risks to the payloads and spacecraft. The operation of multi-instrument payloads is also likely to be complex and therefore increases risk.

3. Management risk—A principal investigator or project manager who is not experienced in satellite engineering and technology can pose a significant risk to the mission. A principal investigator might offer an outstanding scientific mission using low-risk technologies, but he or she might not have the experience to manage the mission properly. Moreover, small teams can introduce risk into the program if there are insufficient checks built into the management approach, if testing is sacrificed, or if the team becomes overworked and burned out. Institution-led and larger programs can experience similar management risks; the failure of the Mars Climate Observer mission may be an example of this. On the other hand, the severe downsizing in recent years of NASA's personnel in technical areas has also significantly diminished its ability to provide the management skills, incisive reviews, and guidance required by current missions.

4. Risk in mission operations—The large number of current and planned science missions is not well matched to the declining financial support for the Deep Space Network and thus imposes risk on mission operations.[52]

PROBLEMS WITH PAST MISSIONS

While the committee has noted several benefits of the FBC approach and linked improvements to this approach, not all missions using it have been successful. A short review of the failures may uncover lessons that will help to ensure the success of future missions.

The failures among missions using the FBC approach include the Mars Polar Lander (MPL), MCO, WIRE, the Lewis and Clark missions, and TERRIERS. Although FBC was not necessarily the main cause of any of these failures, it may have been a significant contributing factor in several cases.

The Lewis mission, launched on August 23, 1997, was intended to demonstrate advanced science instruments and spacecraft technologies for measuring changes in Earth's land surface. Shortly after launch, the spacecraft entered a flat spin that resulted in a loss of solar power and a fatal battery discharge. The Lewis Spacecraft Mission Failure Investigation Board found that the spacecraft was lost as a direct result of the implementation of a technically flawed safe mode in the attitude control system, compounded by the limited control and monitoring of the spacecraft after launch, a decision apparently made to "avoid exhausting the crew."[53] The board also found several contributing indirect causes. More complete documentation, engineering model development, and program reviews might have prevented the failure.

The roles of government and industry have changed significantly in the new FBC era. The Lewis Spacecraft Mission Failure Investigation Board endorsed the concept of FBC in space programs and believed that the new paradigm could be implemented with "sound engineering, and attentive, and effective, management." It noted that the changing roles of government and industry must be planned for and maintained. The Clark mission was terminated in February 1998 owing to escalating mission costs, launch schedule delays, and concerns about the capabilities of the satellite. Originally scheduled for launch in mid-1996, it was part of NASA's Small Satellite

---

[51]Sarsfield, *Cosmos*, 1999, Appendix D.
[52]Space Studies Board, National Research Council, *A Scientific Rationale for Mobility in Planetary Environments*, National Academy Press, Washington, D.C., 1999, p. 55.
[53]"Faster, Cheaper Strategy on Trial," *Science*, 278 (November 14, 1997), p. 1216.

Technology Initiative program. The spacecraft was to include a high-resolution optical element with stereo imaging capabilities that would gather useful environmental data. The instrument development costs and schedule grew unacceptably large.

WIRE was launched March 4, 1999, aboard a Pegasus XL launch vehicle. WIRE's mission was to conduct infrared observations of astronomical objects. WIRE failed because an incorrectly designed electronics box prematurely fired explosive devices, causing early ejection of the instrument's telescope cover.[54] The WIRE cryogen was subsequently quickly expended. The electronics box design had not been peer-reviewed and other system reviews did not focus on analyzing it.

TERRIERS, a part of the Student Explorer Demonstration Initiative (STEDI), was launched in May 1999 to study how the ionosphere affects global telecommunications systems. The satellite was built and launched in 4 years for approximately $4 million. A wiring error caused the spacecraft's attitude control to malfunction and its solar arrays to point away from the Sun. Without solar energy, the spacecraft lost battery power and its ability to operate. The problem, which was not identified during testing, resulted from a design or human error.

The Mars Climate Observer, one of several missions in a series aimed at exploring the planet Mars and its climate history, was lost on September 23, 1999. This occurred as the spacecraft was about to go into orbit around the planet. The Mars Climate Observer Mishap Investigation Board concluded that "the root cause for the loss of the MCO spacecraft was the failure to use metric units in the coding of a ground software file."[55] This problem caused the orbiter to fire its thrusters at levels that put it on an inaccurate trajectory toward the planet. However, the report revealed other underlying errors and problems that led to the failure, including "inattention, miscommunication, and overconfidence," as well as insufficient staffing for the MCO navigation team and other management problems at the Jet Propulsion Laboratory, which managed the project, and at its industrial partner.[56] Although the Investigation Board did not specifically review the impact of FBC practices on the failure, the MCO loss revealed that the processes for ensuring mission success were stressed by the attempts to meet schedule and cost. In addition, "the overall mission stayed within the budget constraints of NASA's 'better, faster, cheaper' mandate . . . but only by using up contingency funds that could have been devoted to other tasks."[57]

More recently, on December 3, 1999, the Mars Polar Lander mission failed to successfully land and deploy on the Martian surface. Because telemetry was purposely planned to be interrupted during the last stage of the landing sequence, very little is currently known about how or why the MPL (and the two separate penetrator probes) failed to function properly from the surface. The failure review board has yet to report on this latest mission failure, but the MPL problem is having a profound effect on the Mars exploration program, its future direction, and, indeed, on how the entire FBC approach is viewed.

In addition to responding to the results of the MPL failure review whenever they are reported, NASA may benefit from further exploring the implications of the Mars mission failures for the direction of the program. Is the Mars program committed to a technology path that is proving to be riskier than its proponents originally anticipated? Are recent losses skewing the program toward sample return missions that lack the critical precursors recommended in planetary science strategy reports? How seriously have the scientific rationale and robustness of the Mars program been affected by the information gaps introduced by recent mission failures? Do current and future mission programs have ample time and budgets to integrate lessons learned from previous failures? Other consequences of the MCO and MPL failures are the additional review processes that missions being developed under FBC principles may be required to heed.[58] The costs of completing these reviews may have to be paid for by reducing the scope of the science, which would be an unfortunate result.

---

[54]NASA, WIRE Mishap Investigation Board Report, 1999.
[55]Mars Climate Observer Mishap Investigation Board, *Phase I Report*, 1999, p. 6.
[56]Hotz, "String of Missteps on Doomed Orbiter," 1999.
[57]Ibid.
[58]"NASA Review Leaves Projects on Launch Pad," *Nature*, 403 (February 10, 2000), p. 583.

While it may not be possible to attribute any of these failures directly to the faster-better-cheaper paradigm, the Lewis, MCO, MPL, and WIRE cases, in particular, raise questions about whether the risks of FBC and the misapplication of its principles played a significant role. An in-depth, independent investigation of the Mars program failures is under way. It will provide further insight into and analysis of FBC principles and will show where the paradigm needs to be modified. The failures of some missions conducted under the FBC approach can be countered by the successes of others, so it will be important to strengthen effective practices while carefully making corrections, where necessary, in how the approach is administered.

2

Science Priorities and NASA Mission Plans

INTRODUCTION

In Chapter 1, the ad hoc committee considered a number of principles and issues to illustrate the complexity of mission choices and the size options that flow from those principles. In this chapter, the committee examines the relationship between the goals established by the science community and the strategic mission plans developed by NASA.

The committee's methodology for this examination was to obtain input from the Space Studies Board (SSB) discipline committees[1] based on their particular perspectives, experience, and previously published analyses. Each discipline committee was asked to answer four key questions:

- Are there arguments for having a spectrum of mission sizes to achieve near-term (10 years) and far-term (10 to 20 years) goals in your discipline?
- What are examples of existing, planned, or proposed missions in that spectrum?
- What criteria would you develop for evaluating the mix of missions you chose?
- Applying the criteria you developed to NASA's portfolio of missions in the NASA strategic plan for your discipline, what are your observations?

The request to the discipline committees for information (Appendix C) also included questions on such factors as the impact of new technology and international cooperation to ascertain the range of mission sizes that scientists consider appropriate for their discipline area and its high-priority scientific objectives.

The ad hoc committee drew on the discipline committees' contributions (Appendix E) in assessing the relationship between NASA's strategic plan, the scientific strategies laid out by the community, and the portfolio of missions assembled to meet the scientific objectives.[2]

---

[1] The Committee on Astronomy and Astrophysics (CAA), the Committee on Earth Studies (CES), the Committee on Planetary and Lunar Exploration (COMPLEX), the Committee on Solar and Space Physics (CSSP), and the Committee on International Space Programs (CISP).

[2] The tables in Appendix E list nonexhaustive examples of missions in the discipline areas and include the mission parameters or top-level scientific objectives for the mission, the relative size range (according to NASA definitions), the mission status of the program, and the timescale of the observations or measurements to be taken.

Chapter 2 is organized into two parts. The first part looks at the linkage between discipline science objectives and agency strategic plans in terms of common, cross-cutting themes that affect the plans for and the development of the Earth science and space science programs. The cross-cutting themes include issues such as portfolio balance, science objectives and larger missions, spacecraft and instrument availability, long-term planning, international partnerships, and the trade-off between more frequent flights for scientific spacecraft and the risk that science objectives will not be met. The second part of the chapter addresses the history, requirements, and programmatic directions of the various disciplines as those factors influence the mission size mix in the disciplines (Earth science, planetary science, solar and space physics, and astronomy and astrophysics).

CROSS-CUTTING THEMES

Is There a Balance of Mission Sizes in Space Science and Earth Science Programs?

Underlying the committee's assessment of mission size trade-offs for NASA's Earth and space science programs is the question of whether the mix of sizes among the ongoing and planned missions addresses high-priority science questions and how the mix might differ by discipline. The committee evaluated balance on two counts: (1) a nonexhaustive account of NASA's ongoing and planned mission mix in each discipline, provided in Appendix E, and (2) the SSB discipline committees' comments on balance in the programs and how well the mission mix responds to high-priority science objectives.

Summarized below are the mission sizes[3] of ongoing and planned missions in NASA's Earth science, planetary science, solar and space physics, and astronomy and astrophysics programs. Attempts to count or quantify the mix of missions sizes raise many questions. Should the count include foreign-led missions in which NASA participates? Should the size of a NASA-led international mission be based on the total cost or just the NASA contribution? Should failed missions be included? A simple count of the nonexhaustive lists shown in Appendix E (not including foreign-led or commercial missions) shows that astronomy and astrophysics has 6 small, 5 medium, and 12 large missions; planetary sciences has 1 small, 6 medium, and 6 large; solar and space physics has 7 small, 9 medium, and 3 large; and Earth sciences has 2 small, 3 medium, and 10 large. Attempts to assess the mission-size mix on gross counts alone, however, overlook several issues, which are detailed below.

The portfolio of mission sizes in astronomy and astrophysics includes large-scale missions that undertake broad areas of astronomy and astrophysics research and small missions (e.g., those in the Explorer line) that address very focused objectives. The medium missions, with the exception of the Gamma Ray Large Area Space Telescope (GLAST), are on the lower end of the range, so the portfolio is weighted to both large and small missions. The scientific objectives for which medium-class missions are appropriate will not be met adequately by the current or planned portfolio laid out by NASA. However, as noted in Chapter 1, international astronomy and astrophysics missions often fit into the medium class and give the discipline a better balance overall.

Unlike NASA's plans for astronomy and astrophysics, which favor small and large missions, its plans for the planetary science program emphasize medium and larger platforms, which can accommodate the power and fuel resources needed to travel to other planetary systems, as discussed elsewhere in this report. While small and medium missions have greater resiliency and flexibility, some high-priority science questions genuinely require large missions (e.g., those where samples must be returned). However, as noted later in this chapter, in the planetary sciences subsection "Discipline-Specific Issues and Concerns," missions with numerous and comprehensive objectives are being planned more or less as medium-size missions (e.g., Pluto Kuiper Express). If the constraint on mission size is too severe relative to the scientific objectives, the inevitable increase in risk may threaten mission success. Thus, the balance of mission sizes for planetary science remains questionable.

In solar and space physics, the SSB has recognized that "although the Explorers do an excellent job of focusing on specific scientific objectives, most of the broader top-priority objectives summarized in the NRC

---

[3]For the purpose of this study, NASA defined mission sizes as small (less than $150 million), medium ($150 million to $350 million), and large (more than $350 million). Costs include expenses for launch and science analysis.

Science Strategy[4] report can only be accomplished with larger, more scientifically capable missions."[5] The SSB also noted that "space physics thus has a critical need for an external line of Solar-Terrestrial Probes (such as TIMED), together with occasional use of larger Frontier Probes, to carry out its science program. The Explorer program can be successful only in such a context."[6]

For several years, there has been a line in the NASA budget for the development of Solar-Terrestrial Probes (STPs), which account for most of the medium-size solar and space physics missions noted in Appendix E that are in the development or definition phases. The continuing need for large missions to carry out science objectives is reflected in extensions to several ongoing large missions, the development of Cluster II, and the recommendation for the Interstellar Probe mission. Solar and space physics plans do not identify any future small missions, as these will evolve from the Explorer program. Small solar and space physics missions must continue to emerge regularly from the Explorer program to sustain portfolio balance at the smaller end.

The Earth science program is in transition. The three most prominent missions (Terra, Aqua, and Chem) are large, and they have been under development for many years. In general, their successors and subsequent ESE missions will be smaller, requiring much less time to define, design, and implement. Thus, a census of sizes at this time of transition does not reflect the true state of NASA's Earth science mission portfolio. The situation is complicated by the inclusion of operational weather satellites, which are by tradition large satellites. Their evolution to smaller systems, should that occur, would reduce development costs while maintaining continuity of observations. In contrast to the quantitative assessment in Appendix E, which shows many larger missions, a qualitative assessment of new starts in NASA's Earth science portfolio would show more medium and smaller missions.

Some Important Science Objectives Require Larger Missions

Inserting smaller missions into the Earth and space science programs promises greater flexibility and more timely science. Focused, well-constrained science goals tend to be good candidates for small missions. Conversely, comprehensive, wide-ranging science goals often demand medium or large missions. Critical scientific objectives that call for missions at the large end of the spectrum can be identified in all four disciplines: planetary sciences, astronomy and astrophysics, solar and space physics, and Earth sciences.

In the planetary sciences, such high-priority objectives as comet nucleus sample return[7] require large missions. In addition, plans to return samples from Mars, such as the Mars Sample-Return Lander 2, would necessitate a large, complex mission, including a lander and rover to collect samples and a Mars return orbiter to carry the samples back to Earth.[8] Exploring for the presence of liquid water on Europa would also call for a larger

---

[4]See Space Studies Board, National Research Council, *A Science Strategy for Space Physics*, National Academy Press, Washington, D.C., 1995.

[5]Space Studies Board, National Research Council, *Scientific Assessment of NASA's SMEX-MIDEX Space Physics Mission Selections*, National Academy Press, Washington, D.C., 1997, p. 14.

[6]Ibid.

[7]". . . COMPLEX's Integrated Strategy report assigns its highest priority to the study of cometary nuclei, which ultimately will require a returned sample. Any sample return is an ambitious task, and previous plans to achieve this objective have been well outside the scope of a small [or medium] mission. COMPLEX's Integrated Strategy also identified the outer solar system (particularly, Neptune and Pluto/Charon) as the key to several questions about solar system origin and evolution . . . Missions to the outer solar system will, however require powerful launch vehicles and specialized power and communications systems. Therefore, unless these requirements are reduced as a result of technological innovation (e.g., development of new propulsion systems), small [and medium] missions are not likely to contribute to this area of planetary science. . . . Even if it does prove feasible to investigate the outer solar system through a small[/medium]-mission program, it may not be cost-effective—that is, the use of small[-medium] missions does not assure that the most science will be returned per dollar spent, especially in the outer solar system. Because of the long flight times and different mission requirements (e.g., long-lived power sources and powerful transmitters) for spacecraft sent to the outer solar system, significant overall economy frequently can be achieved by maximizing the scientific return of any such mission." Space Studies Board (SSB), National Research Council, *The Role of Small Missions in Planetary and Lunar Exploration*, National Academy Press, Washington, D.C., 1995, p. 14.

[8] Space Studies Board, "Assessment of NASA's Mars Exploration Architecture," letter from Ronald Greeley, Chair, Committee on Planetary and Lunar Exploration (COMPLEX), and Claude Canizares, Chair, Space Studies Board, to Carl Pilcher, science program director, Solar System Exploration, National Aeronautics and Space Administration, November 11, 1998, p. 16.

spacecraft to accommodate a suite of possible mobile platforms, including a surface rover, a multifunctional arm on a rover or lander, drilling and coring devices, devices for collecting samples, and a small submarine.[9]

In astronomy and astrophysics many primary science questions require observations by telescopes with pointing systems that distinguish radiation from regions of the sky of small angular extent and observations of objects that are intrinsically faint. Many studies therefore impose relatively strict requirements on spacecraft pointing stability, telescope aperture, and exposure time. These requirements are very difficult or impossible to meet with a small spacecraft using present technology or even technology that may become available in the near term. For example, planet detection and the study of stellar motions with the Space Interferometry Mission (SIM) requires optical interferometry, and the study of high-redshift galaxies with the Next Generation Space Telescope (NGST) requires near-infrared capability. Astronomy and astrophysics programs could be at risk if they are caught between the marginal adequacy of small missions and the cost and time delays associated with observatory-class missions. An intermediate-scale mission range may bridge that gap.

In solar and space physics, the NASA strategic plan and the new Roadmap plan[10] offer mainly medium-size missions (the Solar-Terrestrial Probes, capped at $250 million) and occasional large missions (the Frontier Probes, at more than $250 million). Frontier Probes are more challenging missions and will exceed $350 million or more in total cost. They would address scientific objectives that require difficult orbits—for example, an orbit needed to make in situ measurements of interstellar space—or would involve studies of planetary environments requiring long-duration travel and significant power and fuel resources.

In Earth sciences, large platforms often have been used for measurements that require instruments with large aperture sizes. This is especially true of microwave instruments such as radiometers or synthetic aperture radars (see Chapter 3), whose aperture size is set by fundamental limits (see Chapter 1). For example, several science goals identified in the NRC report *Global Environmental Change: Research Pathways for the Next Decade*[11] require synthetic aperture radar (SAR) measurements. While technological advances may reduce the cost and size of certain electronic subsystems, SAR antenna size is determined primarily by physical limits. Thus, even when a major SAR design goal is smallness, SAR antenna size cannot be reduced to fit that objective.[12] Any mission that relies on SAR to conduct science measurements will continue to require a medium-sized or large platform.

Although some science goals may require large spacecraft, the science community's efforts to focus on only the most essential goals is important. Eliminating hardware that does not contribute to those essential goals may reduce costs and optimize mission size.

Spacecraft and Instrument Availability Affects the Success of FBC

The availability of off-the-shelf spacecraft buses and instruments that rely on existing, flight-proven technologies can help shorten mission development times and keep mission costs from growing. To that end, NASA has started a catalog of available bus designs and is promoting its use, because it can be a cost-effective resource. For example, most of the 20 respondents to the 1998 ESSP program opportunity proposed using a catalog bus.

The benefits of using flight-proven scientific instruments (as opposed to spacecraft buses and subsystems) are more variable. Within specific classes of missions the availability of off-the-shelf hardware has clearly had a large impact on both the missions that are planned and the missions that are selected. Many of the small and medium-size planetary missions, i.e., those in the Discovery and Mars Surveyor mission lines, make extensive use of off-the-shelf instruments, particularly flight spares and copies of existing instruments. For example, seven of the nine instruments on the first three Mars Surveyor orbiters are flight spares from the Mars Observer program. Similarly,

---

[9]Space Studies Board, National Research Council, *A Scientific Rationale for Mobility in Planetary Environments*, National Academy Press, Washington, D.C., 1999, pp. 23-25.

[10]NASA, Office of Space Science, *Sun-Earth Connection Roadmap: Strategic Planning for 2000-2025*, 1999.

[11]National Research Council, *Global Environmental Change: Research Pathways for the Next Decade*, National Academy Press, Washington, D.C., 1999.

[12]Space Studies Board, National Research Council, *Development and Application of Small Spaceborne Synthetic Aperture Radars*, National Academy Press, Washington, D.C., 1998.

the camera system on Stardust (a Discovery mission) makes use of flight spares from Voyager. Clementine's low cost was enabled by the availability of hardware previously developed under the Strategic Defense Initiative (SDI) program.

In the solar and space physics program, the competition for a small number of Explorer mission opportunities demands that proposed science plans be robust, novel, and scientifically world class. The tight cost caps on the programs (University Class Explorer (UNEX), SMEX, and MIDEX) lead researchers to include as much off-the-shelf hardware as possible. In addition to the cost savings from using off-the-shelf hardware and technology, this practice gives proposals an advantage when they undergo a technology evaluation. Such evaluations are typically quite conservative and may rate low-risk flight heritage (existing, proven technology) above science requirements that demand new technologies. Thus, proposals for Explorer missions rely heavily on off-the-shelf equipment to meet cost constraints and to survive the selection process.[13] In solar and space physics, for example, three SMEX selections—SAMPEX, Fast Auroral Snapshot Explorer (FAST), and Transition Region and Coronal Explorer (TRACE)—depended heavily on heritage, and the MIDEX Imager for Magnetopause-to-Aurora Global Exploration (IMAGE) benefited directly and significantly from the closely related mission definition studies funded over the years immediately preceding its selection.[14]

In the short term, instruments that rely on existing technologies may enable a principal investigator to meet rapid schedules and cost constraints. In some cases, available technologies may satisfy science requirements and provide the best option for achieving new science. However, science tends to progress most rapidly when it can take advantage of the greatly increased capabilities enabled by new technologies.[15] Moreover, smaller, shorter-duration missions are a natural arena for opening new observing and measurement windows that require instrumentation having no flight heritage. Absent a new source of technology and new instrument developments, the FBC approach could wither. The choice of spacecraft buses or instrument technologies from a stock list should not be imposed in ways that would discourage the development of new instruments, new subsystems, or new mission concepts.

Long-Term Science Objectives Require Long-Term Planning

Smaller, low-cost missions and their implied constraints on the duration of scientific measurements or observations make long-term planning essential for responding to the needs of long-term and interdisciplinary science. For example, the earlier and larger EOS missions, once part of a 15-year planning horizon, now appear as 5-year, single-satellite versions. Although the measurement requirements for decade-long observations have not changed, the planning outlook is shorter and includes repeated opportunities for missions such as ESSP for 3 to 5 year periods. Similarly, many Sun-Earth Connection missions must study phenomena over a substantial part of the 11-year solar cycle. From a science point of view, it is essential to provide a long-term science plan and to align new mission opportunities to be consistent with that plan. From a technology point of view, it is sensible and feasible to stress the longer duration and greater reliability of a mission as well as smaller size and lower costs. A

---

[13]For more information on the Explorer selection process and potential pitfalls for proposers, see American Geophysical Union, *SPA Section Newsletter* VII(12), February 2, 2000.

[14]SSB, *Scientific Assessment of NASA SMEX-MIDEX Space Physics Mission Selections*, 1997.

[15]Contrast the above to the more traditional approach to executing large, complex missions. First, the scientific objectives of traditional missions had to be clearly defined, analyzed, and agreed to by a sizeable portion of the scientific community. Second, the payload instruments had to be developed to meet those scientific objectives. Third, the platform had to be designed to suit the payload instruments being developed. Finally, the spacecraft had to be integrated and tested to ensure full compliance with all the scientific objectives. If a mission like this is designed to meet a complex series of scientific objectives, it is highly unlikely that a specific payload instrument can be found among existing commercial or government sources that meets those objectives.

The first order of business in traditional missions is development of the scientific instruments required to meet the ambitious scientific objectives. These specialized instruments often take a long time to develop and may be subject to schedule delays and overruns that impact the entire mission. Nevertheless, complex missions typically yield extraordinary scientific results, satisfy a broad scientific constituency, and also leave a heritage of technologies and instrumentation developments that benefits future missions.

management strategy that incorporates scientific vision and emphasizes long-term planning is important for facilitating the balance between long-duration science missions and more frequent, shorter-duration missions.

International Partnerships Help to Balance Mission Sizes in a Portfolio

In the section on international cooperation in Chapter 1, the committee discussed how the FBC approach can affect international collaboration. This section discusses how such cooperation can contribute to the mix of mission sizes in NASA's portfolio of Earth and space science missions.

Cooperative international programs are a clear example of how flexibility in mission planning has produced excellent science at low costs. Many of the large missions (e.g., Cassini and UARS) would not be possible in anything resembling their present form had they not had international contributions. The ability to forge low-cost (to the U.S. taxpayer) international collaborations has sometimes meant that the United States has been able to avoid duplicative missions. For example, because ROSAT, a German-U.S.-U.K. astrophysics mission, and ASCA, a Japanese-U.S. mission, were the natural successor missions to Einstein (with an increase in sensitivity, resolution, and bandpass by a factor of between 3 and 5), the United States was able to leapfrog directly to Chandra, a Great Observatory.

In space science, ESA has a well-defined, long-term plan called Horizon 2000 and Horizon 2000 Plus. Although its long-term plan is less well defined, Japan has a plan for the next 5 years. In some areas of endeavor (such as cosmic microwave background studies), the ESA mission (Planck) is larger and more sophisticated than the analogous U.S. mission (Microwave Anisotropy Probe, (MAP)) but will fly about 6 years later. In other areas of astrophysics (such as the Far Infrared and Submillimeter Telescope (FIRST) and the International Gamma-Ray Astrophysics Laboratory (INTEGRAL)), ESA's missions have diminished the pressure for a U.S.-only mission because ESA's goals address some of the same goals as the NASA strategic plan.[16] ESA and the United States are cooperating closely in the CLUSTER-II space physics mission, with ESA having the lead role. Similarly, NASA and the Japanese have a very successful ongoing collaboration in space physics with the Geotail mission and in solar physics with the Yohkoh and the Solar-B missions; Japan's Institute of Space and Astronautical Science is the lead agency for all three collaborations.

In the Earth sciences, international contributions have enhanced many missions but have introduced discontinuities in data for others. The Tropical Rainfall Mapping Mission (TRMM) and most of the successful ESSP projects would not have been approved by NASA if there had not been substantial foreign involvement. Many U.S. researchers recognize the importance of radar altimetry for oceanic science, global weather forecasting, and the observation of long-term climate signals. France has been particularly important for precision radar altimetry, an area on which NASA is not currently focusing. The U.S.-French radar altimetry mission, TOPEX/Poseidon, continues to collect important data on sea-surface measurements (see Box 1.1 in Chapter 1). France is taking the lead role in the follow-up to TOPEX/Poseidon, Jason-1, which is slated for launch in 2001. Similarly, NASA has chosen not to focus on synthetic aperture radar measurements, an area in which Europe, Japan, and Canada have been leading.

An apparent characteristic of foreign missions is that many fit within (or approach) the medium-cost window of U.S. missions, as defined in Appendix B. As a result, the gap created by current NASA plans, at least with

---

[16]The flight of the first very long baseline interferometry mission (HALCA) by the Japanese seems to have affected U.S. plans for a similar experiment and has stimulated future U.S. proposals. A space-based radio interferometry mission, Advanced Radio Interferometry between Space and Earth (ARISE), to be overseen by an international advisory group, is described as one of the possible new missions in the NASA strategic plan.

The impact of the longer-range ESA Horizon 2000 Plus plan will not be known until early 2000, after the revision of the NASA strategic plan. The ESA plan includes GAIA, an advanced astrometric mission; the infrared space interferometry mission DARWIN, whose goals include the first direct detection of terrestrial planets in orbit around stars other than our Sun and the first high-spatial-resolution imaging in the 6-μ to 30-μ wavelength region; LISA, a gravitational wave experiment with close U.S. collaboration; and XEUS, a very-large-area X-ray imaging and spectroscopic mission. Since DARWIN overlaps with TPF, LISA has strong U.S. science community support, and since the science goals for XEUS overlap (but are much more ambitious than) those of Constellation-X, the committee anticipates that ESA plans will have a substantial impact on the NASA program. An approved ESA mission has never been canceled.

respect to space-based astronomy, is somewhat filled. Similarly, in the planetary sciences, foreign-led missions are providing the larger platforms required for key scientific objectives that involve sample return. One example is Rosetta, a cometary mission that includes 12 instruments from several nations, among them the United States. It will rendezvous with comet Wirtanen in 2012. The mission also includes a lander to conduct in situ measurements of the comet's nucleus. With NASA's cancellation of the Comet Rendezvous Asteroid Flyby (CRAF) mission and the Champollion lander, Rosetta becomes a critical link in addressing the SSB's most important near-term scientific priority, which is to explore comets and other primitive solar system bodies.[17]

International partnerships—both those led by NASA and those led by foreign partners—have been very successful in solar and space physics. NASA is the lead agency for the Global Geospace Science (GGS) program, which includes the Wind and Polar spacecraft, both of which have important international components. Conversely, ESA is the lead agency for Ulysses, the Solar and Heliospheric Observatory (SOHO), and Cluster II, and Japan is the lead for Yohkoh, Geotail, and Solar B—all examples of foreign missions with high scientific priorities and significant U.S. participation. The current and planned programs in solar and space physics continue to include foreign participation, which contributes to the NASA strategic plan in these disciplines.

Despite the fact that international missions clearly contribute to a balance in the mission mix needed to address high-priority science goals, there is no way, at present, of including these missions in NASA's long-term planning effort. Indeed, the current U.S. strategic planning process is limited in its ability to coordinate with other space agencies on existing and planned international missions and to include in the planning information about the follow-up to its own missions.

Trade-off: Does More Frequent Science Mean Greater Risk?

In addition to the management and technical risks that the FBC approach may pose for missions (Chapter 1, section on risk), the FBC paradigm can also threaten NASA's ability to carry out its strategic plans and to address the priorities established by the scientific community.

In astronomy and astrophysics, notwithstanding the success of such astronomy missions as SWAS and the Far Ultraviolet Spectroscopic Explorer (FUSE), the failure of the High Energy Transient Explorer (HETE)—owing to launch vehicle problems—and WIRE and the partial failure of ALEXIS indicate that the risks associated with small missions can be quite high. Several smaller, shorter-duration missions in the astronomy and astrophysics program await completion and launch (e.g., MAP, the Full Sky Astrometric Mapping Explorer (FAME), HETE-2, the Galaxy Evolution Explorer (GALEX), Swift, and the Cosmic Hot Interstellar Plasma Spectrometer (CHIPS)). If they succeed, they will justify flying more frequent missions at higher risk, and the scientific return will have been great indeed. However, if a substantial fraction of the smaller, shorter-duration missions are either partial or total failures, then the FBC concept will not have succeeded.

In solar and space physics the principal experience with FBC includes the notable successes with recent Explorers (e.g., the three SMEX missions: SAMPEX, FAST, and TRACE) and the Student Explorer Demonstration Initiative (STEDI) mission SNOE. In contrast, another STEDI mission, TERRIERS, failed (as discussed in Chapter 1), and the Thermosphere-Ionosphere-Mesosphere Energetics and Dynamics (TIMED) mission faced budget cuts that led to a descoping of the science objectives, downsizing of the mission, and subsequent delays. Thus far—as long as the problems with STEDI and TIMED do not recur—missions aimed at providing more frequent science cannot be judged to have introduced significant risk into the solar and space physics program.

In the planetary science program, however, the recent loss of the Mars Climate Observer and the Mars Polar Lander exemplified the potential risks that the FBC approach poses for the strategic plan. Earlier, in January 1999, the Near Earth Asteroid Rendezvous (NEAR) satellite failed to complete the first objective of its mission when the

---

[17]As conceived in the mid-1990s, Rosetta was to carry two landers, the NASA-supplied Champollion and the Franco-German RoLand. However, a mismatch of schedules between NASA and ESA forced the deletion of Champollion from the Rosetta mission. NASA later restructured its comet-exploration efforts and resurrected Champollion as a stand-alone spacecraft under the aegis of the New Millennium technology demonstration program. Called Space Technology 4 (previously Deep Space-4), the new program was finally canceled for budgetary reasons in 1999.

spacecraft did not execute a crucial engine burn required for rendezvous with the Eros asteroid. Unlike the Mars missions, NEAR had another opportunity to complete its original objectives and successfully executed a rendezvous with Eros on February 14, 2000. The loss of science from the Mars failures will probably affect the feasibility of the originally planned scientific investigations as well as the mission designs for the next Mars missions.

Another dimension of risk lies in the mission selection process and the potential for Explorer AOs, which are segregated by size and cost (e.g., MIDEX only or SMEX only), to encourage overambitious science plans. The science objectives may be squeezed (or expanded) to fit the AOs since researchers may want to take as much advantage as possible of each flight opportunity. The committee wondered if AOs could be designed to enable mission size and cost to be better matched to the proposed science. NASA might wish to examine this approach.

DISCIPLINE-SPECIFIC ISSUES AND CONCERNS

The issues described in the first part of this chapter are important because they affect all space-based science programs. However, each discipline also has a unique set of questions or observing requirements that must be understood on its own. This section outlines discipline-specific issues for the Earth science, planetary science, solar and space physics, and astronomy and astrophysics fields.

Earth Sciences

Earth science includes diverse scientific disciplines such as oceanography, land processes, atmospheric sciences, meteorology, climate, and geodesy, all of which utilize observations and measurements from space. For more than 30 years these disciplines have relied heavily on observations from meteorological satellites that now include NOAA's polar-orbiting operational environmental satellites (POES) and its geostationary operational environmental satellites (GOES), the Defense Meteorological Satellite Program (DMSP) polar-orbiting satellites, and their foreign counterparts. The U.S. operational polar-orbiting series is being converged into the National Polar-Orbiting Operational Environmental Satellite System (NPOESS), which is preparing to begin on-orbit operations about 2009. The fact that there are both research and operational Earth-observing satellites is significant for the present discussion because the future uses of these satellites will have to address both research and operational objectives.[18]

Mission Mix and NASA Plans

For many years, mission scope and spacecraft size for space-based Earth science research grew to accommodate the increased size and capability of instruments. However, the budgets that accompanied large spacecraft such as those in EOS were cut and the program was redesigned significantly. In an effort to maintain program efficiencies, as well as to sustain Earth science observations over a longer time horizon, the revised EOS plan, as it appeared in 1995, depended on successive reflights of substantially similar systems. That plan suffered under budget pressure, leading to the cancellation of the second and third reflights of large satellites in the EOS plan, but the accompanying insertion of smaller satellites solved some of the budgetary and technical problems.[19] (Box 2.1 shows examples of large and small Earth science missions.) One consequence, however, is that the Earth Science Enterprise (ESE) Science Implementation Plan does not respond to some of the criteria for evaluating the mix of

---

[18]The integration of critical climate research measurements into operational missions such as NPOESS 1 entails risks for collecting these measurements. This is the subject of the SSB report *Issues in the Integration of Research and Operational Satellite Systems for Climate Research: I. Science and Design*, National Academy Press, Washington, D.C., 2000. Both NASA and NOAA are making diligent efforts to accommodate such risks because missions like these integrate research and operational capabilities for collecting long-term climate data.

[19]See Space Studies Board, National Research Council, *Earth Observations from Space: History, Promise, and Reality*, National Academy Press, Washington, D.C., 1995.

BOX 2.1
Earth Science Accomplishments with Large and Small Missions

UARS

One of NASA's flagship Earth science missions, the Upper Atmosphere Research Satellite (UARS), requested by Congress in 1976 and launched in 1991, continues to provide scientists with a better understanding of the linked physical and chemical dynamics of the stratosphere.[1] Weighing 6 metric tons and carrying 10 scientific payloads, the UARS spacecraft was large in both scale and cost (roughly $750 million in life-cycle costs).

UARS is an international endeavor with contributions from France, the United Kingdom, and Canada. It has contributed directly to international research on changes in the chemistry of the atmosphere, including ozone depletion due to NO_x and chlorofluorocarbons.[2]

QuikSCAT

Launched 8 years after UARS, in June 1999, the NASA Quick Scatterometer (QuikSCAT) gathers all-weather, high-resolution measurements of near-surface winds over Earth's oceans. Named for its quick replacement of the NASA Scatterometer carried on the lost Japanese Advanced Earth Observation Satellite (ADEOS) mission, QuikSCAT uses a spare spacecraft bus and instrument. Small in comparison to UARS, it weighs 970 kg and was developed in a faster-better-cheaper mode at a cost of approximately $95 million (life-cycle costs). The satellite carries the SeaWinds instrument, a sophisticated microwave radar that measures wind speed and direction.[3,4]

GRACE

Planned for launch on June 23, 2001, from Plesetsk, Russia, the Gravity Recovery and Climate Experiment (GRACE), which consists of 380-kg twin satellites, will begin its mission to create a new model of the variations in Earth's gravity field. The first mission in an FBC program known as the Earth System Science Pathfinder (ESSP), GRACE will map Earth's gravity fields with unprecedented accuracy by measuring the distance between the two satellites using the Global Positioning System and a microwave ranging system.

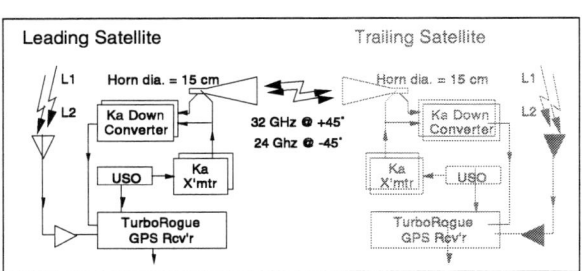

GRACE is an $85.9 million (for the U.S. contribution) project that also has extensive participation from Germany.[5]

---

[1] Space Studies Board, National Research Council, and European Science Foundation, *U.S.-European Collaboration in Space Science*, National Academy Press, Washington, D.C., 1998, pp. 83-84.
[2] Image source: <http://uarsfot08.gsfc.nasa.gov>.
[3] See <http://winds.jpl.nasa.gov/missions/quikscat/quikindex.html>.
[4] Image source: <http://winds.jpl.nasa.gov/missions/quikscat/quikindex.html>.
[5] Image source: <http://essp.gsfc.nasa.gov/grace/index.html>.

mission sizes listed in Appendix E.[20] In addition, the time line for planning follow-up missions in the EOS program has been shortened. Sustaining critical measurements over many years is more complicated technically and programatically.

The mix of missions sizes in NASA's current plan provides less assurance that the Earth science program may be relied upon to answer fundamental and long-term science questions.[21] Furthermore, important questions now surround long-term shifts in the coupled Earth system (the boundaries or interactions between ocean and land, land and atmosphere, ocean and atmosphere, and people and the natural world), and the answers to such questions may be useful for policy makers. Smaller, shorter-duration missions make such interdisciplinary investigations more difficult programmatically, because it is difficult to ensure that the data sets have sufficient overlap and duration to answer these questions.[22,23]

The need to blend different sources of Earth science data is another critical factor affecting decisions on the portfolio of mission size in NASA's Earth science program. Certain science goals endorsed by the science community, such as measuring the sea-level rise, verifying the stabilization of atmospheric ozone depletion, and detecting an increase or decrease in the mass of polar ice, require long time horizons, good data continuity, and well-characterized and well-calibrated instruments. To separate short-term variability in these phenomena from long-term trends often requires a blend of satellite, in situ, and model data. In each case, the mission design from concept to implementation must be in tune with the scientific objectives. Data from smaller or shorter missions as well as larger or longer ones can contribute to valuable long-term measurement programs if well planned and executed. Thus, instrumentation sufficient to support a long-term science objective does not necessarily imply many years of large annual expenditures.

In addition, the ESE Technology Development Plan[24] should be encouraged, but not to the exclusion of the science themes. NASA is to be commended for the recently introduced Instrument Incubator Program (IIP), aimed at bringing newly proposed measurement techniques from the stage of concept to the stage where they can compete for a space flight opportunity. However, it is difficult to plan and implement the transitioning of research instrumentation developed under very tight schedules to operational satellite systems, which are designed to incorporate new technology more slowly.

Planetary Sciences

The four principal scientific goals identified in the SSB's planetary strategy are intensive studies of (1) comets, (2) Mars, and (3) the Jupiter system, and (4) the search for extrasolar planets.[25] Each goal is addressed within NASA's strategic plan. The medium and small missions of the Discovery line have enabled the community to address a wide variety of other science goals (e.g., studies of the Moon by Lunar Prospector and of asteroids by NEAR and Deep Space-1 (a New Millennium mission)). Nevertheless, the vast distances from Earth, extreme

---

[20]High on the list of problems that were meant to be solved by the U.S. Global Change Research Program (USGRP) were the need for well-calibrated observations, the need to maintain critical observations, and the need for a focused scientific strategy. ESE's use of smaller, shorter missions and the pressure to off-load continuing observations onto systems of the operational agencies runs counter to those needs.

[21]Space Studies Board, "Report of the Task Group on Assessment of NASA's Plans for Post-2002 Earth Observing Missions," letter from Claude R. Canizares, Chair, Space Studies Board; Marvin A. Geller, Chair, Task Group on Assessment of NASA's Plans for Post-2002 Earth Observing Missions; Eric J. Barron and James R. Mahoney, Co-chairs, Board on Atmospheric Sciences and Climate, and Edward A. Frieman, Chair, Board on Sustainable Development, to Ghassem Asrar, associate administrator, Earth Science Enterprise, NASA, April 8, 1999.

[22]NASA, *Earth Science Enterprise, Earth Science Implementation Plan, Version 1.0*, April 1999, p. 61.

[23]For example, a large part of the Earth science data that address several of the science goals identified in the GCRP priorities highlighted in the *Pathways* report is provided by SAR satellites. Save for NASA's Seasat (1978), NASA has participated in SAR satellite programs only peripherally through international agreements with Canada, the European Space Agency, and Japan, whose successful satellite programs have led the field for the past 20 years. NASA has flown several SARs on one-week demonstration shuttle missions, but the impact of brief in-space operations on many Earth science questions is minimal. Although the cost of any realistic SAR satellite would place it in the medium to large category, the data would serve several high-priority science objectives.

[24]NASA, *Earth Science Enterprise Technology Development Plan*, 1999.

[25]Space Studies Board, National Research Council, *An Integrated Strategy for the Planetary Sciences: 1995-2010*, National Academy Press, Washington, D.C., 1994.

differences in (and poor knowledge of) operating environments, and long trip times make planetary missions inherently more complex and, hence, more expensive than missions that are flown closer to Earth.

The growth in spacecraft size, number of instruments, mission complexity, and duration all contributed to the escalating costs of planetary missions in the late 1980s (typified by Mars Observer and Cassini).[26] The concomitant risk associated with large, infrequent planetary missions was a major impetus for the FBC approach. Specifically, after the failure of Mars Observer in 1993, the risks in Mars exploration were managed by launching two spacecraft at every opportunity (which happened to be every 26 months) within a mission line of $145 million in FY98, $228 million in FY99, and $250 million in FY00 (although this plan was not sufficient to eliminate risk, as evidenced by the MCO and MPL failures). The creation of the Discovery class of missions, whose mission objectives are selected via an open competition, allowed a rapid response to new ideas for scientific exploration. However, care must be taken when comparing a small mission such as Lunar Prospector (~$70 million) with a large mission such as Cassini (~$2.5 billion). If Cassini were to be repeated with current technology and an FBC approach, the costs would be much less (how much less would depend on the degree of risk the developers would be willing to bear), but it would still be a large, complex mission. Cassini's comprehensive set of 12 instruments and the Huygen's probe allows a systematic exploration, including simultaneous measurements, of the complexities of Saturn's atmosphere, rings, and moons over several years. While the five instruments on Lunar Prospector have provided useful information about the composition of the lunar surface and gravity field, the potential scientific accomplishments of Cassini could not be achieved with a series of multiple missions of the Lunar Prospector class. Thus, solar system exploration is optimized by a mixed portfolio of mission sizes, which includes occasional large missions for major objectives (particularly necessary in the outer solar system), many medium-size missions, and a few small missions for focused targets (in, say, a 1:10:5 ratio).[27] (Box 2.2 shows examples of large and small planetary science missions.)

Mission Mix and NASA Plans

The Discovery line of missions (initiated in NASA's FY94 budget) is intended to collect data to answer questions about the solar system within a total life-cycle cost of $300 million per mission (including launch). The initial skepticism over the scientific value of such smaller, shorter-duration missions as Mars Pathfinder, NEAR, Lunar Prospector, and Clementine has been moderated by the valuable and sometimes unexpected science return from these missions. They have, for example, contributed to a better understanding of the presence of hydrogen at the Moon's poles (Lunar Prospector); the small-scale geology and chemical composition of rocks on Mars (Mars Pathfinder); the magnetic stripes on Mars (Mars Global Surveyor); and the low density of the asteroid Mathilde (NEAR).

The FBC approach clearly has many benefits, such as flexibility, frequency, and timeliness of missions, as well as controlling excessive growth of mission costs and development time.[28] However, tight constraints on mission costs can result in underestimating the funds necessary to return and analyze the data (e.g., Lunar Prospector (see Box 2.2)); reliance on spare instruments from previous missions (e.g., Stardust);[29] reliance on nonmission funds (e.g., PIDDP)[30] to support instrument development (most missions);[31] or dependence on supplemental funds to pay for the launch.

---

[26]SSB, *The Role of Small Missions*, 1995, p. 4.

[27]SSB, *Small Missions*, 1995, p. 14; SSB, *Integrated Strategy*, 1994, pp. 182-183.

[28]SSB, *Integrated Strategy*, 1994, p. 30; SSB, *Small Missions*, 1995, p. 15.

[29]Stardust was launched in 1999 to collect material from a comet and interstellar dust and return it to Earth for analysis.

[30]"The Planetary Instrument Definition and Development Program supports the advancement of spacecraft-based instrument technology that shows promise for use in scientific investigations on future planetary missions. The goal of the program is to define and develop instruments or instrument components to the point where the instruments may be proposed in response to future announcements of flight opportunity without additional extensive technology development." See "Research Opportunities in Space Science 1998," NRA 98-OSS-03, issued February 5, 1998, Appendix A 3.5 Planetary Instrument and Development Program, available at: <http://spacescience.nasa.gov/nra/98-oss-03>.

[31]SSB, *Small Missions*, 1995, p. 19.

BOX 2.2
Planetary Science Accomplishments with Large and Small Missions

Cassini

Cassini, launched October 15, 1997, is a mission to study Saturn's atmosphere, magnetic field, rings, and moons. The 2,160-kg spacecraft is a joint NASA, European Space Agency (ESA), and Italian Space Agency endeavor. The Cassini spacecraft, including the orbiter's 12 instruments and the 6 instruments on the Huygens probe, is one of the largest, heaviest, and most complex interplanetary spacecraft ever built, having cost $2.55 billion (including operations and science analysis).

Its scientific objectives are to conduct the following observations:

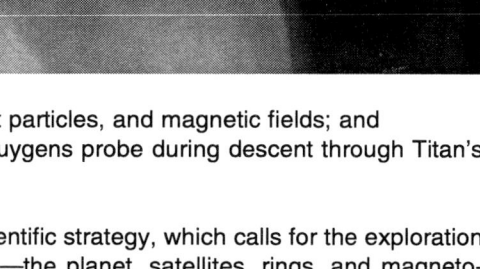

- Orbital remote sensing of Saturn's atmosphere, icy satellites, and rings;
- In situ orbital measurements of charged particles, dust particles, and magnetic fields; and
- Detailed measurements with six instruments on the Huygens probe during descent through Titan's dense nitrogen atmosphere to the surface.[1]

These science objectives respond directly to the NRC scientific strategy, which calls for the exploration of the outer planets, including an intensive study of Saturn—the planet, satellites, rings, and magnetosphere—as one its highest priorities. Cassini is expected to reach Saturn in 2004 and begin its 4-year primary orbiter mission.[2,3]

Lunar Prospector

Lunar Prospector is the first in a class of planetary probes, the Discovery line of missions. Developed at a total cost of $68 million, it was launched on January 6, 1998, beginning its 5-day trip to the Moon, where it remained in orbit for 18 months. The 300-kg spacecraft was equipped with a gamma-ray spectrometer, a neutron spectrometer, a magnetometer-electron reflectometer, an alpha-particle spectrometer, and equipment for a Doppler gravity experiment. The critical scientific objectives of the Lunar Prospector were as follows:

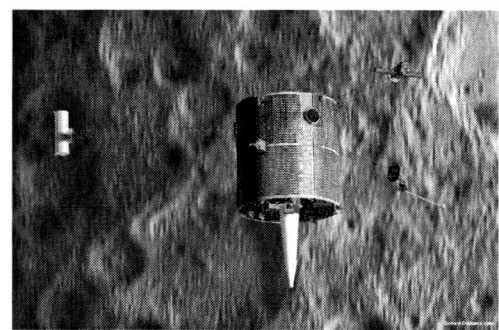

- To prospect the lunar crust and atmosphere for potential resources, including minerals, water ice, and certain gases;
- To map the Moon's gravitational and magnetic fields; and
- To learn more about the size and content of the Moon's core.

continued

BOX 2.2 Continued

Lunar Prospector data were used to develop the first precise gravity map of the entire lunar surface and the first global maps of the Moon's elemental composition, two primary scientific objectives recognized in an earlier NRC report, *Strategy for Exploration of the Inner Planets, 1977-1987*.[4] In addition, despite the fact that the Moon's magnetic field is relatively weak, the Lunar Prospector was able to confirm the presence of local magnetic fields. In a final attempt to detect water on the Moon, on July 31, 1999, it was crashed into a crater near the south pole of the Moon. However, no signature of water was detected.[5,6]

NEAR

On February 17, 1996, the Near Earth Asteroid Rendezvous (NEAR) mission was launched to make the first quantitative and comprehensive measurements of an asteroid's dimensions. The primary scientific goals are to assess the following:

- Bulk properties: size, shape, volume mass, gravity field, and spin state;
- Surface properties: elemental and mineral composition, geology, morphology, and texture; and
- Internal properties: mass distribution and magnetic field.[7]

The 805-kg spacecraft is managed by the Johns Hopkins University Applied Physics Laboratory and is the first launch of NASA's Discovery program, an initiative for small planetary missions with a maximum 3-year development cycle and a cost cap of $224 million (life-cycle costs). Despite earlier complications, on February 14, 2000, NEAR rendezvoused with Eros, a large near-Earth asteroid, inserted itself into orbit around Eros, and began the year-long mission.[8]

---

[1] SSB and ESF, *U.S.-European Collaboration*, 1998.
[2] SSB, *Integrated Strategy*, 1994; Space Studies Board, "On the Scientific Viability of a Restructured CRAF Science Payload," letter from Space Studies Board Chair Louis J. Lanzerotti and Committee on Planetary and Lunar Exploration Chair Larry W. Esposito to Lennard A. Fisk, associate administrator for NASA's Office of Space Science and Applications, August 10, 1990; Space Studies Board, "On the CRAF/Cassini Mission," letter from Space Studies Board Chair Louis J. Lanzerotti, transmitting a report of the Committee on Planetary and Lunar Exploration to Lennard A. Fisk, associate administrator for NASA's Office of Space Science and Applications, March 30, 1992.
[3] Image source: Painting by Michael Carroll, available electronically at <http://www.jpl.nasa.gov/cgi-bin/gs?/cassini/moreinfo/pix/dropoff.jpg>.
[4] SSB, *Strategy for Exploration of the Inner Planets: 1977-1987*, National Academy of Sciences, Washington, D.C., 1978.
[5] A.B. Binder, 1998. "Lunar Prospector: Overview," *Science* 281:1480-1484; Space Science Board, National Research Council, *Strategy for Exploration of the Inner Planets: 1977-1987*, National Academy of Sciences, Washington, D.C., 1978; Space Science Board, National Research Council, *A Strategy for Exploration of the Outer Planets: 1986-1996*, National Academy Press, Washington, D.C., 1986; SSB and ESF, *U.S.-European Collaboration*, 1998.
[6] Image source: <http://george.arc.nasa.gov/dx/basket/storiesetc/lpcrapix.html>.
[7] Space Studies Board, National Research Council, *The Exploration of Near-Earth Objects*, National Academy Press, Washington, D.C., 1998.
[8] Image source: <http://near.jhuapl.edu/NEAR/images/near2.gif>.

Achieving Major Science Objectives with a Series of Missions

When major science objectives can be implemented by using a series of small and medium-size missions, there are recognizable benefits. As an example, replacing previously large-scale missions (e.g., Mars Observer and Mars sample return) with a series of small and medium missions has improved mission resiliency and provided an opportunity to address new scientific questions. Splitting up the original payload from the lost Mars Observer mission into three missions (Mars Global Surveyor, the now-failed Mars Climate Observer, and the Mars Surveyor 2001 orbiter) not only addressed the scientific objectives of Mars Observer but also provided an opportunity to fly three new instruments. Similarly, the redundancy inherent in the current Mars sample-return architecture enables multiple samples to be collected by a variety of means at widely separated sites. In addition, the use of two Mars ascent vehicles improves the likelihood that at least one set of samples will be returned to Earth. Nevertheless, the series of Mars missions will not accomplish the scientific goals set by NASA if the mission architecture relies heavily on developing new technology under tight schedules, constrained costs, and without means to recover from failures.

Squeezing Large Science Objectives onto Medium-Size Missions

Important scientific objectives such as sample returns, surface landers, and flights to the outer solar system cannot be achieved without larger missions. Nevertheless, as noted previously, a number of important scientific goals best addressed by large missions are being implemented as medium-sized (or nearly so) missions. This has led to programmatic problems (e.g., cost overruns and schedule delays) that may have contributed to the cancellation of ST-4/Champollion and an overreliance on alternative sources of funding to develop instruments and spacecraft technologies. These risks may be most evident in the Outer Planets program, which includes the Europa Orbiter and the Pluto/Kuiper Express missions as the first in a series of probes to explore organic-rich environments. Pluto/Kuiper Express (full life-cycle costs estimated at $354 million) and Europa Orbiter (full life-cycle costs estimated at $460 million) are characterized as large, according to the size categories set for this study. Both missions are technically demanding in their need to reach the outer parts of the solar system. Many researchers question, however, whether the allocated budgets will be sufficient to meet both the technical challenges and the science objectives, or whether the science objectives will be cut severely in order to stay within budget.

As a consequence of promising to do more with less, there has been a serious reduction not only in the number of scientific instruments per mission but also in the fractional mass of the scientific payload on planetary missions (e.g., Europa Orbiter and Pluto/Kuiper Express).[32] While there is clearly value in limiting the tendency to always want to attach just a few more instruments onto a spacecraft, there comes a point when there are too few instruments and too little scientific return to justify the cost of getting there. The key is to optimize the scientific return with a mission sized just large enough to adequately address the priority scientific objectives.

Independent of mission size, squeezed budgets tend to affect mainly the later parts of a mission (the analysis, synthesis, and presentation of data), which then severely limits the mission's scientific value[33] (as happened, for instance, with Magellan, Galileo, Clementine, Lunar Prospector, Mars Global Surveyor, and all aspects of the International Solar-Terrestrial Physics program).

---

[32]The mass of scientific instruments as a fraction of the total spacecraft mass (including propellant) is typically 11 percent for Discovery-size missions (Clementine, NEAR, Mars Global Surveyor). Missions to the outer solar system require large propellant masses (particularly for orbiters) and so tend to have lower payload mass fractions. Moreover, the science payload mass fraction for outer solar system missions has been decreasing with time: 13 percent for Voyager, 6.8 percent for Cassini, 2.5 percent for Pluto/Kuiper Express, and 2 percent for Europa Orbiter. See Sarsfield, *Cosmos;* D. Matson and J.-P. Lebreton, "The Cassini/Huygens Mission to the Saturian System," *Space Science Reviews*, in press.

[33]SSB, *Small Missions*, 1995, pp. 22-23.

Solar and Space Physics

Solar and space physics are the primary disciplines involved in NASA's Sun-Earth Connection (SEC) theme in the Office of Space Science (OSS). The studies of the Sun are examining its interior, surface, atmosphere, magnetic field, and solar variability in all of its manifestations. The solar atmospheric studies extend outwards to include the solar wind and the heliosphere and are also studying the physical interactions between the Sun and planetary environments, particularly Earth's. The study of the space environment of Earth encompasses the magnetosphere, the ionosphere, and the upper atmosphere; coupling and energy transfer processes into, out of, and within the magnetosphere; and the various effects of solar-induced activity throughout geospace. The study of the magnetosphere and ionosphere is not restricted to Earth but extends to other planets, comets, and bodies in the solar system. There are complementary, and at times partially overlapping, interests in space physics and planetary science, and this has been appropriately reflected in the scientific objectives and accomplishments of several major missions through the years, such as Pioneer, Voyager, Galileo, and Cassini.

Mission Mix and NASA Plans

The slate of recommended missions in the new SEC Roadmap[34] builds on, and is generally consistent with, the OSS Strategic Plan of 1997 and with the Committee on Solar and Space Physics' assessment of that plan.[35] Thus, each STP and Frontier Probe contributes significantly to one or more of the scientific objectives of the OSS Strategic Plan. The OSS Strategic Plan of 1997 does not specifically address mission cost or cost caps, but it does support the FBC approach, as discussed in the plan's appendix on metrics.[36] The new SEC Roadmap incorporates a strategy to implement the plan's science objectives primarily using STP missions (less than $250 million), which are medium-size, and occasionally by using Frontier Probes (greater than $250 million), which are medium-size to large. The Roadmap also adopted the FBC approach by adhering to cost constraints in the mission planning process itself, which meant that the mission plans accepted for the Roadmap were ones that met cost-cap requirements after evaluation by the Roadmap technology teams.

The Roadmap does not explicitly address whether a recommended mission fell short of optimal scientific objectives because of the set cost categories. However, in all likelihood some of the STPs will find it very difficult to meet optimal science objectives without exceeding the $250 million cost cap. The mission definition team for the Solar Terrestrial Relations Observatory (STEREO) has evidently already exceeded the funding ceiling, and the team for the Magnetospheric Multi Scale (MMS) is struggling to finalize mission designs and objectives that will keep the mission within the budget ceiling. An exciting and truly novel mission such as Magnetospheric Constellation, which would consist of 100 or so spacecraft networked in orbit, would undoubtedly have to make scientific compromises to meet the cost constraints of an STP mission.

Scientific Priorities and the Mission Portfolio

In the new SEC Roadmap, scientific priorities are couched in the form of quests and campaigns. Quests are major questions for which answers are needed to understand solar variability and its effects on the solar system and life on Earth. Implementation is achieved through organized campaigns. Quests address the following:

- Why does the Sun vary?
- How do Earth and the planets respond to solar variability?

---

[34] NASA, *Sun-Earth Connection Roadmap: Strategic Planning for 2000-2025*, 1999.

[35] See NASA, *The Space Science Enterprise Strategic Plan: Origin, Evolution, and Destiny of the Cosmos and Life*, November 1997; Space Studies Board, National Research Council, *An Assessment of the Solar and Space Physics Aspects of NASA's Space Science Enterprise Strategic Plan*, National Academy Press, Washington, D.C., 1997; NASA, *Sun-Earth Connection Roadmap: Strategic Planning for 2000-2025*, 1999.

[36] See NASA, *The Space Science Enterprise Strategic Plan*, 1997, Appendix B.

- How do the Sun and the galaxy interact?
- How does solar variability affect life and society?

Campaigns address the following:

- The origins of solar variability;
- The effects of solar variability on the corona and the solar wind;
- The geospace environment;
- Comparative planetary space environments;
- The heliospheric boundary and nearby galactic environment; and
- Space weather.

Each recommended mission in the solar and space physics program must make significant contributions to one or more of the SEC campaigns. The order of the mission queue is based on the following:

- Relative importance of the scientific objectives;
- Likelihood of achieving the scientific objectives by the mission plan;
- Potential for discovery and understanding;
- Breadth of the science (does it contribute to more than one campaign?);
- Urgency and relevance to society;
- Programmatic issues: timeliness (do it now or later?);
- Budgetary impact;
- Technology readiness; and
- Development costs and capability.

All recommended missions are required to have an approved plan for education and outreach in accord with SEC and NASA expectations. SEC missions are designed to have direct relevance to other NASA science themes and the interests of other government agencies.

In space and solar physics there is a clear need for a portfolio of mission sizes. Certain focused scientific objectives can often be accomplished with small or medium-size missions, which could be conducted through the Explorer or STP programs. Other scientifically more challenging and complex objectives would require far greater financial resources and large missions to be successful. Some of the Frontier Probes would be large missions. (Box 2.3 shows examples of one large mission, SOHO, and two small space physics missions, SAMPEX and TRACE.) An implementation plan might have a number of reasons for including a large mission, depending on the science to be accomplished. Some of the conditions that could lead to large missions are a long observation time line (solar variations or sunspot cycle effects); multiple instruments of high resolution (microphysics of particles and fields); highly stable platforms (for remote observations); vast physical parameter ranges in the operating environment (heliospheric observations outwards to interstellar space); interdisciplinary missions (planetary missions to investigate both the planet and its environment); and use of multiple spacecraft or constellations (to separate spatial and temporal effects or to make complementary observations simultaneously).

Astronomy and Astrophysics

Astrophysical sources radiate at all wavelengths of the electromagnetic spectrum, yet Earth's atmosphere is transparent in only a few windows. Even at visible wavelengths accessible to ground-based telescopes, there are gains in angular resolution, dynamic range, and astrometric precision that are achievable only from space. Space astronomy is thus essential for progress across the whole field of astrophysical research.

BOX 2.3
Space and Solar Physics Accomplishments with Large and Small Missions

SOHO

"The most comprehensive space mission ever devoted to the study of the Sun and the heliosphere,"[1] the Solar and Heliospheric Observatory (SOHO) is a joint international project between the European Space Agency and NASA. Measurements and images taken from SOHO are helping scientists better understand the structure and dynamics of the solar interior using helioseismology techniques. In addition, researchers analyzing SOHO data are gaining insight into the physical processes that form and heat the Sun's corona and into the solar wind and its acceleration processes.

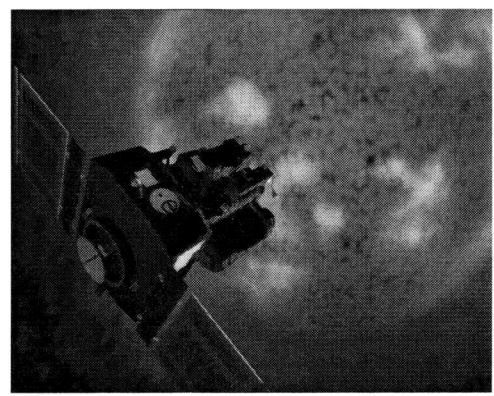

The 1,850-kg spacecraft is equipped with 12 instruments, including helioseismology instruments to study the structure and dynamics of the solar interior from the deep core to the outermost layers and remote-sensing instruments, including extreme ultraviolet and ultraviolet imagers, spectrographs, and coronagraphs to view the outer solar atmosphere and corona.

Since SOHO began its 3-year mission in 1995, the observatory has provided the first image of the convection zone of a star; the first tracing of the slow-speed solar wind near the equatorial current sheet; the first detection of elements and isotopes in the solar wind; and the first observations of coronal mass ejections (CMEs) that generated subsequent disturbances. These disturbances were observed by other spacecraft to establish a cause-and-effect relationship for a solar system event that extended from the Sun to the solar wind to Earth's magnetosphere and ionosphere. Having completed its original mission, SOHO was extended another 5 years, to 2003.[2,3]

SAMPEX

Space physicists have a more complete understanding of the highly energetic, charged particles emanating from the magnetosphere and cosmic rays around the Sun and Earth as a result of the Solar Anomalous and Magnetospheric Particle Explorer (SAMPEX). This research includes particles trapped in Earth's magnetosphere and those that enter the magnetosphere from interplanetary space. In particular, SAMPEX studied the composition and charge state of anomalous cosmic rays, which are not of solar, galactic, or extragalactic origin.

Launched in 1992, weighing 170 kg, and costing $80 million (life-cycle costs), this FBC-style satellite has contributed to the fundamental understanding of anomalous cosmic rays in interplanetary space, a high-priority goal identified in previous NRC strategy reports.[4]

continued

BOX 2.3 Continued

The SAMPEX satellite carries a set of four detectors designed with a high resolution and sensitivity to sense anomalous, galactic, and magnetospheric energetic particles. Data collected from the SAMPEX instruments have shown definitively that anomalous cosmic rays are mostly singly ionized and that upper atmosphere NO_x changes with the level of flux of the precipitating energetic electrons. SAMPEX is currently in an extended mission phase.[5,6]

TRACE

At 224 kg, the Transition Region and Coronal Explorer (TRACE), launched in 1998 and costing approximately $72 million (life-cycle costs), is contributing to scientific understanding of the processes that lead to solar variability. It is providing continuous observations of the Sun at the ultraviolet and extreme ultraviolet wavelengths. These observations are taken from a single, high-resolution telescope.[7] The images and observations made from TRACE provide insight into the three-dimensional magnetic structure emanating from the Sun and help define the geometry and dynamics of the upper solar atmosphere, known as the transition region and corona.

In addition, the telescope is acquiring solar images taken through filters that select different spectral features. By comparing the temporal evolution of events as seen through different filters, investigators are gaining critical information about the origin and evolution of local energy-release processes and the rearrangement of coronal structures such as coronal holes.[8]

---

[1] SSB and ESF, *U.S.-European Collaboration*, 1998, p. 51.
[2] SSB, *SMEX-MIDEX*, 1997, pp. 5-6.
[3] Image source: <http://sohowww.nascom.nasa.gov/gallery/SC>.
[4] SSB, *A Science Strategy for Space Physics*, 1995.
[5] SSB, *SMEX-MIDEX*, 1997.
[6] Image source: <http://lepsam.gsfc.nasa.gov/www/public/sampex.html>.
[7] SSB, *SMEX-MIDEX*, 1997, p. 9.
[8] Image source: <http://vestige.lmsl.com/TRACE/Public/Gallery/Images/>.

Mission Mix and NASA Plans

Scientific merit is the first criterion, as noted in Appendix E, for evaluating the balance in a portfolio of missions. The goals of the space astronomy and astrophysics program outlined in the NRC report *A New Strategy for Space Astronomy and Astrophysics*[37] involve obtaining answers to a set of fundamental questions about planets, star formation, and the interstellar medium; stars and stellar evolution; galaxies and stellar systems; and cosmology and fundamental physics. (Examples of large and small astronomy and astrophysics missions and their contributions to the field are shown in Box 2.4.)

The highest-priority goals include the following, ranked according to their priority:

- Determination of the geometry and content of the universe by measuring the fine-scale anisotropy of the cosmic microwave background radiation;
  - Investigation of galaxies near the time of their formation at very high redshift;
  - Detection and study of planets around nearby stars; and
  - Measurement of the properties of black holes of all sizes.

Other important, unranked goals include the following:

- Study of star formation by, for example, high-resolution far-infrared and submillimeter observations of protostars, protoplanetary disks, and outflows;
  - Study of the origin and evolution of the elements;
  - Resolution of the mystery of the cosmic gamma ray bursts; and
  - Determination of the amount, distribution, and nature of the dark matter in the universe.

Scientific Priorities and the Mission Portfolio

In mapping the scientific goals to the portfolio of missions, the first-ranked goal, which involves studies of the fine-scale anisotropy of the cosmic background radiation, can be accomplished with a small mission such as MAP. The second-ranked goal, the study of galaxies near the time of their formation, requires a large-aperture infrared telescope that must be housed on a large platform designed to accommodate telescope size and weight (NGST). One of the unranked scientific goals, resolution of the mystery of the cosmic gamma-ray bursts, can be pursued with a small mission such as HETE-2 and a medium-size mission such as Swift. Other goals, such as studying the origin of the elements and measuring the properties of black holes, can be done well with missions in the medium cost range.

Since 1994, when NASA adopted the FBC paradigm for conducting missions, the agency has selected an excellent set of small and medium missions. However, the fact that very few of these had actually flown as of early 2000 means that it is difficult to decide if the change in emphasis to smaller missions has been successful.[38] In the 10-year period from 1998 to 2007, NASA is planning for 15 UNEX missions, 10 SMEX missions, and 8 MIDEX missions to be shared among three themes: the Sun-Earth Connection, the Astronomical Search for Origins, and the Structure and Evolution of the Universe. NASA has also pursued a more vigorous program with its somewhat

---

[37] Space Studies Board, National Research Council, *A New Science Strategy for Space Astronomy and Astrophysics*, National Academy Press, Washington, D.C., 1997.

[38] The Bahcall committee (formally known as the Astronomy and Astrophysics Survey committee, which authored the NRC report *The Decade of Discovery in Astronomy and Astrophyics*, 1991) recommended that the number of Explorers in the medium-cost category be increased to six. Ongoing medium-cost missions such as ACE and FUSE were completed. GLAST, which might be considered medium, is in the planning stages. In general, however, the astronomy and astrophysics portfolio has contained few medium-cost missions. Swift and FAME, which are also in the planning stage, are in the lower range of the medium category. See also Space Studies Board, "On ESA's FIRST and Planck Missions," letter to Wesley T. Huntress, Jr., NASA Associate Administrator for Space Science, from Claude R. Canizares, Chair, Space Studies Board, and Robert Dynes, Chair, Board on Physics and Astronomy, February 18, 1998.

BOX 2.4
Astronomy and Astrophysics Accomplishments with Large and Small Missions

Hubble Space Telescope

Ranked as the highest priority in the 1970s astronomy decadal survey, the "Greenstein report,"[1] the Hubble Space Telescope was launched in April 1990. It was the first of the Great Observatories designed for sensitive, high-angular-resolution observations in the ultraviolet through near-infrared spectral range. It features a suite of instruments that are upgraded by periodic shuttle missions. It is the largest orbiting observatory ever built; the total mission cost since its inception has been over $8 billion (to NASA, life-cycle costs including use of the shuttle) and about $550 million (to ESA).

Since the refurbishment mission that corrected the spherical aberration induced by the 2.4-m primary mirror, the telescope has delivered images with a sharpness close to the limit imposed by diffraction. These images are significantly sharper than those delivered by ground-based telescopes, revealing entirely new phenomena at smaller physical scales. The mission has made crucial contributions across the whole of astrophysics, from planets (impact of comet Shoemaker-Levy 9 with Jupiter) to candidate supermassive black holes (nuclear regions of galaxies) to cosmic evolution (morphological structures in the most distant known galaxies).

The general outcome is that the Hubble Space Telescope has had the greatest impact of any observatory-type facility available in space.[2,3]

Submillimeter Wave Astronomy Satellite

The Submillimeter Wave Astronomy Satellite (SWAS) is a small Explorer, launched in December of 1998 and designed to study the chemical composition of interstellar gas clouds. Its primary objective is to survey water, molecular oxygen, carbon, and isotopic carbon monoxide emissions in a variety of star-forming regions in the Milky Way. The spacecraft is making detailed 1 degree x 1 degree maps of these species in giant molecular and dark cloud cores with an angular resolution of 4 arcminutes.

The overall goal of the mission is to gain a greater understanding of star formation by determining the composition of interstellar clouds and by establishing the means by which these clouds cool as they collapse to form stars and planets. Other SWAS targets include external galaxies, circumstellar envelopes, planetary nebulae, and solar system objects (e.g., water features in Jupiter and in comets).

The spacecraft spent its first viewing year exploring the Milky Way through a number of targets. It will return to its original orbit, from which it operates in an observation mode, and make more detailed studies of selected targets.[4,5]

---

[1] National Research Council, Astronomy Survey Committee, *Astronomy and Astrophysics for the 1970s*, National Academy Press, Washington, D.C., 1972.
[2] SSB and ESF, *U.S.-European Collaboration*, 1998, p. 44.
[3] Image source: <http://www.stsci.edu/hst>.
[4] See <http://cfa-www.harvard.edu/cfa/oir/research/swas.html>.
[5] Image source: <http://cfa-www.harvard.edu/cfa/oir/research/swas.html>.

larger and more capable Delta-class Explorers. A recent example of a Delta-class Explorer is the FUSE mission, which was launched in June 1999. At a total cost of $204 million, FUSE would be classified as a medium-size mission.

NASA's current flight mission program in space astronomy comprises mostly large new starts (such as SIM, NGST, Constellation X, and Terrestrial Planet Finder (TPF), each with total costs exceeding $550 million) or missions with total costs of less than $140 million. There are few current or planned missions in the $200 to $550 million range. The small number of true medium-sized missions has split science goals into those that are accommodated on small and MIDEX platforms and those that engage the broad community and result in programs such as NGST.

Portfolio and Planning

In addition to increasing the use of smaller missions, the move toward an FBC paradigm has led to changes in the process for choosing missions. In the past, NASA, through its advisory committees, working groups, and external and internal experts, chose the area of science to which a new mission would be devoted (e.g., the AO that resulted in the Rossi X-ray Timing Explorer (RXTE) mission called specifically for an X-ray timing mission). Under present policy, the AO calls for any mission that fits within the cost caps and is consistent with the general guidelines of the AO. The thematic approach to AOs allowed the science community to respond to certain scientific goals in NASA's strategic plans, while the new approach, although it has yielded excellent scientific proposals, does not define scientific areas and therefore cannot be incorporated into long-term plans.

3

Summary and Recommendations

Chapter 3 presents the ad hoc committee's findings in response to the tasks set forth in the original charge and its recommendations to NASA based on the analyses in Chapters 1 and 2. In its deliberations the committee focused on the implications of FBC for the space programs' tolerance for risk; the scope, diversity, and timeliness of the science investigated; the results and analytical products produced from space research missions; the availability and use of advanced technology; the training and educational opportunities for students; and the role of international cooperation in supporting the mix of mission sizes in NASA programs. The committee's findings and recommendations reflect the importance of these factors in facilitating a balanced portfolio of mission sizes for achieving high-priority science for NASA's Earth science and space science programs. In light of the myriad and complex considerations bearing on mission planning, the committee did not prescribe what the mix of mission sizes should be.

THE CHARGE

The request for this report originated from the Senate Appropriations Committee in its FY99 report; NASA commissioned the NRC to conduct the assessment. The charge to the committee sets three key tasks:

1. Evaluate the general strengths and weaknesses of small, medium, and large missions in terms of their potential scientific productivity, responsiveness to evolving opportunities, ability to take advantage of technological progress, and other factors that may be identified during the study;

2. Identify which elements of the SSB and NASA science strategies will require medium or large missions to accomplish high-priority science objectives; and

3. Recommend general principles or criteria for evaluating the mix of mission sizes in Earth and space science programs. The factors to be considered will include not only scientific, technological, and cost trade-offs but also institutional and structural issues pertaining to the vigor of the research community, government-industry-university partnerships, graduate student training, and the like.

STRENGTHS AND WEAKNESSES OF SMALL AND LARGE MISSIONS

Small Missions

Small missions (defined by NASA as missions costing less than $150 million) play a key and compelling role in the space-based Earth and space science programs. They are responsible for decreasing the time to science, leading to scientific analyses that can be conducted within years rather than decades. Smaller programs tend to be flexible and responsive to new scientific opportunities. They also provide for answering the more focused questions that emerge from larger-scale research activities (e.g., the Great Observatories). Shorter development periods for small missions reduce their overall costs and provide additional degrees of budgetary freedom for NASA's Earth and Space Science Enterprises.

What stands out prominently is the professional vitality and community involvement small missions can offer. Researchers who might have waited a lifetime to analyze data are stimulated, freshened, and sharpened when given the opportunity to conduct high-priority, high-quality science in a shorter period. In turn, these programs afford undergraduate and graduate students opportunities to participate in and experience the complex and organic nature of science, from proposal to development, to analysis, to publication. The STEDI program is a case in point, and the desirability of its broadly educated "alumni" is borne out by industry's strong demand for them.

Small satellites are appropriate for highly focused and relatively limited scientific objectives. However, their flexibility, while opportunistic for science, poses challenges for strategic planning and meeting long-term science objectives. The ESSP and Explorer lines rely on open AOs. While these AOs result in excellent scientific proposals, they may not give the scientific community a sense of the overall planning and direction for the program. Moreover, they may make it more difficult for international partners to submit proposals individually or join a proposing team. The balance between planning and flexibility is a fine one. Nonetheless, the committee believes that efforts to provide appropriate planning for small missions will ensure that their scientific contributions enhance the overall Earth and space science programs and the community's objectives.

Large Missions

Smaller missions are not replacements for the scientific scope that large (defined by NASA for this study as missions costing more than $350 million) platforms can accommodate. The strength of large missions lies in their ability to accommodate complex scientific objectives requiring long-term measurements, sophisticated instruments, large mass, and/or instrument and spacecraft redundancy. When technology development and instrument development were included in large-scale programs, as they often were, the programs brought forth a wealth of experience and instrumentation that benefited subsequent programs. The scientific output of missions such as Hubble, Galileo, Magellan, and UARS testifies to their value in terms of scientific achievement.

Current priorities continue to demand large as well as medium and small missions. Long-term records of climate fluctuations, for example, are required before scientists can draw conclusions about global climate trends or predict future impacts. Such long-term observations necessitate spacecraft with sufficient power. They must often assemble a time series of observations spanning decades or more. These requirements may translate into high levels of mass and a large enough platform to accommodate robust subsystems. Mass and platform size are also critical elements for astronomical observations requiring large-aperture telescopes. Looking deeper into the universe or making more accurate spectroscopic assessments of planets around remote stars demands larger, more complex spacecraft systems. The discussion in Chapter 1 of the physical constraints on and principles of conducting space-based science articulates these issues.

Missions that address complex science have required longer development periods, which automatically increases a program's cost. Moreover, as noted in the technology discussion in Chapter 1, the prospect of sending a large, expensive spacecraft carrying several sophisticated instruments to the outer planets reduces tolerance for risk that might be inherent in newer, more capable technologies or leaner management. By the same token the larger size of such a spacecraft could accommodate redundant systems, which is an effective mechanism for managing risk on large spacecraft carrying numerous science payloads and traveling long distances. Thus, an intelligent balancing of capabilities, perceived risk, and available resources will continue to be the principal challenge to NASA managers.

RECOMMENDATIONS IN RESPONSE TO THE CHARGE

Faster-Better-Cheaper Principles

The committee's findings begin with a recommendation on the broader implementation of faster-better-cheaper principles. The committee found that FBC methods of management, technology infusion, and implementation have produced useful improvements regardless of absolute mission size or cost. However, while improvements in administrative procedures have proven their worth in shortening the time to science, experience from mission losses (Mars Climate Observer and Lewis, for example) has shown that great care must be exercised in changing technical management techniques lest mission success be compromised.

Recommendation 1:
Transfer appropriate elements of the faster-better-cheaper management principles to the entire portfolio of space science and Earth science mission sizes and cost ranges and tailor the management approach of each project to the size, complexity, scientific value, and cost of its mission.

Science Goals and Mission Size

In the Earth sciences long-term climate measurements are needed, and many of NASA's research programs will have to be more closely integrated with the nation's operational programs. Operational missions such as those of NOAA require redundancy and continuity of a complex set of measurements. These requirements usually translate into medium-size or large spacecraft. Climate research, on the other hand, requires sustained, accurate, and calibrated measurements. These requirements often translate into a mix of mission sizes at any one time, but they, too, imply a commitment to a long-term measurement strategy. The operational and research measurement variables overlap, but not completely.

For example, long-term weather forecasting and the development of climate computer models place more rigorous demands on the horizontal and vertical resolution for temperature and moisture atmospheric profiles. These data are gathered by infrared and microwave atmospheric sounders on polar-orbiting satellites. Models were satisfied only 10 years ago by data sampled at 4-km vertical intervals and a 250-km spatial grid. Now scientists need 1-km vertical resolution and a spatial grid of less than 20 km. In response, the number of frequency bands in the instruments has had to be increased by a factor of approximately four, and aperture sizes have been increased to attain the smaller grid size. In spite of advances in technology, the newer instruments are larger and more massive than their predecessors and require larger spacecraft.

In the planetary sciences, high-priority questions requiring samples to be returned to Earth from Mars or the core of a comet, exploration to the solar system's outer planets, and planetary or cometary landers would all require large-scale missions.

In solar and space physics, SSB science strategies and NASA strategic plans call for a full portfolio of mission sizes to carry out the scientific objectives of the discipline. The SEC long-range strategy has identified the medium-size and large missions needed for its science plan. Medium-size missions include those with clusters of near-Earth-orbiting spacecraft and certain solar missions with more focused objectives. Large missions are needed where orbital requirements are very severe (such as missions to access and study interstellar space or the polar regions of the Sun), where a long, continuous time line of observations is required (such as to observe solar variations and sunspot cycle effects), and where planetary environments (such as the plasma electrodynamics at Jupiter and Io) are studied.

The astronomy and astrophysics community has implemented large-scale missions and continues to call for several more: SIRTF, SIM, GLAST, TPF, Constellation X, NGST, and Laser Interferometer Space Antenna

(LISA). These missions respond to the scientific imperative to detect the range of a radiation emitted by both common and exotic sources, and it is this imperative that drives the technology. Typically, work at the frontier requires enhanced sensitivity and enhanced angular and spectral resolution, with the consequent need for large missions. For example, studies of galaxies near the time of their formation require sensitive, high-angular-resolution imaging capabilities in the near-infrared part of the spectrum, capabilities that will be offered by the Next Generation Space Telescope. The sheer size of the telescope aperture required, coupled with the low operating temperatures, necessitates a large mission platform.

In considering the role of science goals in planning for a portfolio of mission sizes, the committee found the following:

- The nature of the phenomena to be observed and the technological means of executing such observations are constrained fundamentally by the laws of physics, such that some worthwhile science objectives cannot be met by small satellites.
- The strength and appeal of faster-better-cheaper is to promote efficiency in design and timely execution—shorter time to science—of space missions in comparison with what are perceived as less efficient or more costly traditional methods.
- A mixed portfolio of mission sizes is crucial in virtually all space and Earth science disciplines in order to accomplish a variety of significant research objectives. An emphasis on medium-size missions is currently precluding comprehensive payloads on planetary missions and has tended to discourage planning for large, extensive missions.

Recommendation 2:
Ensure that science objectives—and their relative importance in a given discipline—are the primary determinants of what missions are carried out and their sizes, and ensure that mission planning responds to (1) the link between science priorities and science payload, (2) timeliness in meeting science objectives, and (3) risks associated with the mission.

Technology Development

A further key point is that small missions (and their concomitant short development times) have depended on access to previously developed instruments and technologies. Without a source of new instruments, the missions using faster-better-cheaper principles cannot be sustained. Indeed, smaller missions are intended, to some extent, to tolerate more risk from new instruments and/or technologies. However, to date the selection processes for medium (defined by NASA for this study as $150 million to $350 million) programs such as Discovery have been surprisingly risk-averse.

The committee considered the role of technology as it assessed mission size trade-offs for Earth and space science missions and found the following:

- Technology development is a cornerstone of first-rate Earth and space science programs. Advanced technology for instruments and spacecraft systems and its timely infusion into space research missions are essential for carrying out almost all space missions in each of the disciplines, irrespective of mission size. The fundamental goal of technology infusion is to obtain the highest performance at the lowest cost.
- The scientific program in Earth and space science missions conducted under the FBC approach has been critically dependent on instruments developed in the past. The ongoing development of new scientific instrumentation is essential for sustaining the FBC paradigm.

Recommendation 3:
Maintain a vigorous technology program for the development of advanced spacecraft hardware that will enable a portfolio of missions of varying sizes and complexities.

Recommendation 4:
Develop scientific instrumentation enabling a portfolio of mission sizes, ensuring that funding for such development efforts is augmented and appropriately balanced with space mission line budgets.

Cost of Access to Space

In addition to trade-offs in the areas of management, scientific scope, and technology noted above, several other factors must be taken into account when deciding on mission-size mixes in NASA's space program. Specifically, the committee found that access to space is a primary determinant of timeliness and cost in executing science missions:

- The high cost of access to space remains one of the principal impediments to using the best and most natural mix of small and large spacecraft. While smaller spacecraft might appear to be the right solution for addressing many scientific questions from orbit, present launch costs make them an unfavorable solution from an overall program budgetary standpoint. Moreover, larger missions, too, are plagued by the excessive costs per unit mass for present launch vehicles.
- The national space transportation policy requiring all U.S. government payloads to be launched on vehicles manufactured in the United States prevents taking advantage of low-cost access to space on foreign launch vehicles.

Recommendation 5:
Develop more affordable launch options for gaining access to space, including—possibly—foreign launch vehicles, so that a mixed portfolio of mission sizes becomes a viable approach.

International Collaboration

The committee found that international collaboration has proven to be a reliable and cost-effective means to enhance the scientific return from missions and broaden the portfolio of space missions. Nevertheless, it is sometimes considered, within NASA, to be detrimental, perhaps because it adds complexity and can bring delays to a mission. It is also perceived to give a mission an unfair advantage and, in part, to increase NASA's financial risk.

In the past NASA had within its budgets an international payload line, which was an extremely useful device for funding the planning, proposal preparation, and development and integration of peer-reviewed science instruments selected to fly on foreign-led missions. This line offered the U.S. scientific community highly leveraged access to important new international missions by providing investigators with additional opportunities to fly instruments and retrieve data, especially during long hiatuses between U.S. missions in a given discipline.

Recommendation 6:
Encourage international collaboration in all sizes and classes of missions, so that international missions will be able to fill key niches in NASA's space and Earth science programs. Specifically, restore separate, peer-reviewed announcements of opportunity for enhancements to foreign-led space research missions.

OTHER FINDINGS ON ISSUES AFFECTING MISSION SIZE MIX

The committee's deliberations and findings as reported in Chapters 1 and 2 and above provide the framework for establishing the right balance of small, medium, and large missions in NASA's science enterprise. There is a clear need for a mixed portfolio of mission sizes and scopes; any decisions in this regard must of course be tempered by the budgetary and resource limitations operative at any given time.

The committee examined four additional issues that are important in considering the trade-offs on mission size mix: (1) education, (2) assessment of risk, (3) data analysis, and (4) evaluating the science return from space missions.

Education

The committee notes that the emphasis on education as part of the FBC approach is positive. However, it is not aware of any attempts to assess how the quantity or quality of educational activities varies with mission size. The committee believes it is important to optimize the quantity and quality of educational activities (in line with the class of the mission) associated with all space and Earth science missions (Chapter 1, section "Implementation," subsection "Education").

Risk

The committee found that risk associated with smaller, shorter-duration missions will generally be higher than risk associated with traditional programs. However, such risk can be handled effectively, provided sound management and lessons learned from past mission failures are applied. Risks can be minimized by ensuring that the level and quality of staffing is commensurate with the degree of complexity and risk associated with missions conducted under the FBC approach (Chapter 1, section "Risk").

Data Analysis

The scientific outcomes of a mission include data, data analysis, scientific findings, and publications. The scientific value and return from missions must be considered when evaluating mission scope and scale and the balance of mission sizes. Specifically, the committee found the following:

• In the Earth sciences, research on climate requires data from long-term satellite observations in addition to data collected in situ to identify changes and trends. A mix of mission sizes—including shorter-duration, narrowly focused missions; larger operational platforms; and in situ sources versus remote data collection—intensifies the need for careful planning, coordination, calibration, and integration among data sets.

• Good sensor characterization and calibration, along with continuing data product validation, are essential attributes of space-based measurement systems. Smaller, shorter-duration missions sometimes provide insufficient calibration and validation, which compromises the science return.

• Space research missions are successful only if they extract the optimum scientific value from the data set generated. An appropriate allocation of the investment—between the space system and instrumentation elements; data calibration, characterization, and validation; and the subsequent data analysis effort—is essential to a logical evolution of mission sequences in a given field of Earth or space science. The committee believes it is important to develop an implementation plan for each science mission, regardless of size, that will support data integrity (characterization, calibration, and validation) and scientific analyses beyond the data-acquisition part of the mission (Chapter 1, sections "Fundamental Science Limits" and "Measuring and Enhancing the Scientific Return on the Investment").

Evaluating the Science Return

The committee notes the following:

• Comparing small and large missions after they have achieved their objectives to assess the quality or cost-effectiveness of the science product is inherently complex and not amenable to simple formulas. Peer review that takes place in advance against a background of long-term vision and science planning to establish mission priorities is an effective way to evaluate the scientific potential of a mission and the appropriateness of its size.

• The success of a particular mission can be judged by comparing its accomplishments to the original goals, recognizing that there may be unexpected discoveries and that other benefits may be realized only later.

APPENDIXES

A

Letter of Request from NASA to the Space Studies Board

National Aeronautics and
Space Administration

Headquarters
Washington, DC 20546-0001

APR 22 1999

Reply to Attn of: SR

Professor Claude Canizares
Chair, Space Studies Board
National Academy of Sciences
2101 Constitution Avenue, NW
Washington, DC 20418

Dear Professor Canizares:

Both the Earth Science Enterprise and the Space Science Enterprise have, during the past few years, created near- and longer-term goals for their scientific programs, as well as missions to accomplish those research goals. The Space Studies Board (SSB), as well as other boards and task groups, have had a major role in shaping and reviewing those science plans and planned missions. We believe that the planned missions in each of our two disciplines consist of balanced sets of small, medium, and large missions, tailored to the range of science questions to be addressed, and as dictated by the physics of each observation to be conducted.

Recently, there has been much emphasis on selecting, building, and launching "smaller, faster, cheaper, better" missions. Sometimes this emphasis created the impression that NASA has completely abandoned the medium and larger missions which we have traditionally emphasized. Congress has directed NASA to:

> ". . contract with the National Research Council (NRC) for a study across all space science and Earth science disciplines to identify missions that cannot be accomplished within the parameters imposed by the smaller, faster, cheaper, better regime. The [study] report should focus on the next 15 years, and attempt to quantify the level of funding per project that would be required to meet the specified scientific goals. The report also should identify any criteria

APPENDIX A

and methods that could be used to measure whether the science accomplished using small satellites is better than that accomplished with larger, more complex spacecraft. The report is to be submitted to the Committee no later than September 30, 1999."

The purpose of this letter is to request that the SSB conduct such a study. We are addressing the SSB since it is the NRC entity with the longest connection to the science interests of both Earth and space science disciplines, and assume that other entities will be involved as appropriate.

The level of funding for each project in the next 15 years will depend on its precise science objectives and implementation. Since mission planning for the latter part of the 15-year interval is currently underway in both the Office of Space Science and the Office of Earth Science, and since important elements of these offices' programs will be accomplished through competitive selections in the community-based Explorer, Discovery, and Earth System Science Pathfinder lines, a complete allocation of science objectives to particular missions cannot be made at this time. For the later part of the 15-year period, your assessment may address generic mission ideas presented in the strategic plans.

For the purposes of this study, we request that you consider three broad mission categories: Small missions with a total cost (including launch, operations, and science analysis) of less than $150M; medium missions with total costs up to $350M; and large missions, including strategic missions and observatories, with a total cost (including launch, operations, and science analysis) of more than $350M.

We recognize that the completion date of September 1999 is likely to pose a difficult challenge, and request your views about the earliest feasible time scale for completion of a study addressing the following specific questions:

1. What are the general criteria for assessing strengths and limitations of small, medium, and large missions in terms of scientific productivity, including quality and amount of science value returned, responsiveness to

evolving opportunities, ability to take advantage of technological progress, and other factors?

2. Within the near term, which science goals as provided in our strategic plans will require the use of medium and large missions, in terms of the definitions above and the strengths and limitations in (1) above?

Although, the scientific goals and our mission plans are evolving, fairly complete recent summaries can be found at http://www.earth.nasa.gov/visions/index.html, and at http://spacescience.nasa.gov/strategy/1997. We look forward to receiving your proposal, including study costs and estimated completion date. If you require further information on our current plans, and to discuss study charter and implementation, please work with Dr. Jack Kaye, Director, Research Division, and Ms. Anngienetta Johnson, Director, Program Planning and Development in the Office of Earth Science, and Dr. Guenter Riegler, Director, Research Division, Office of Space Science. For contractual matters, please contact Ms. Dolores Holland at 202/358-0834.

Sincerely,

Ghassem R. Asrar
Associate Administrator
 for Earth Science

Edward J. Weiler
Associate Administrator
 for Space Science

cc:
NRC/Mr. J. Alexander

B

Statement of Task

BACKGROUND

In the past several years there has been much emphasis in NASA on selecting, building, and launching "smaller, faster, cheaper, better" space missions. How NASA implements this new paradigm has ramifications for the relative roles of universities, industry, and NASA centers; opportunities for students to participate in experimental space research; modes of support for advanced technology development; ease of coordination in planning international cooperation; and other programmatic and institutional areas. Sometimes this emphasis creates the impression that NASA has completely abandoned the medium and larger missions it tended to emphasize in the past. Apparently in response to such concerns, Congress directed NASA to "contract with the NRC for a study across all space science and Earth science disciplines to identify missions that cannot be accomplished within the smaller, faster, cheaper, better regime. The [study] report should focus on the next 15 years, and attempt to quantify the level of funding per project that would be required to meet the specified scientific goals. The report also should identify any criteria and methods that could be used to measure whether the science accomplished using small satellites is better than that accomplished with larger, more complex spacecraft."[1]

NASA's smaller, faster, cheaper, better strategy involves efforts to streamline mission development cycles and costs, thereby increasing the number and frequency of flight missions and providing more opportunities for investigators to access spaceflight data. Such missions can be developed and launched in a few years, at a flight rate of more than 10 per year, and at costs of no more than a few hundred million dollars each. In contrast, missions such as Viking, Voyager, Galileo, the Hubble Space Telescope, and AXAF required decades to develop and budgets of up to $1 billion or so. On the other hand, small missions do incur certain scientific costs and risks—for example, when they require compromises in the breadth or depth of measurements that can be accomplished on a small spacecraft or when design practices or features require risk-taking to meet cost constraints. Furthermore, some scientific investigations simply cannot be performed on small spacecraft—for example, when a large telescope aperture is required to gather enough light to conduct the necessary observations. These considerations lead one to conclude that a strong program probably needs a mix of mission sizes. These can be thought of as a

---

[1] U.S. Senate. 1998. Department of Veterans Affairs, Housing and Urban Development, and Independent Agencies Appropriations Bill, 1999, 105th Congress, 2nd Session, S. Report 105-216.

portfolio whose content is determined by many variables, including scientific priorities and yield, cost, frequency of flight, technology utilization, and technical risk.

PLAN

The Space Studies Board will draw on its own membership and that of its standing discipline committees to conduct an independent assessment of how mission size relates to the ability of planned or potential missions to address high-priority scientific goals in Earth and space science over the next 15 years. Mission sizes will be treated in three broad categories—small (total cost less than $150 million), medium, and large (total mission cost greater than $350 million). Recent science strategy reports from the Space Studies Board discipline committees[2] and NASA's space and Earth science strategic plans will be used to define science goals and priorities for the period. Specifically, the board will do the following:

- Evaluate the general strengths and weaknesses of small, medium, and large missions in terms of their potential scientific productivity, responsiveness to evolving opportunities, ability to take advantage of technological progress, and other factors that may be identified during the study;
- Identify which elements of the SSB and NASA science strategies will require medium or large missions to accomplish high-priority science objectives; and
- Recommend general principles or criteria for evaluating the mix of mission sizes in the Earth and space science programs. The factors to be considered will include not only scientific, technological, and cost trade-offs but also institutional and structural issues pertaining to the vigor of the research community, government-industry-university partnerships, graduate student training, and the like.

SCHEDULE

The Space Studies Board will begin work on this project at its June 1999 meeting, continue work during its September executive committee meeting, and conclude report preparation at its November meeting. The relevant standing committees of the board (Committee on Astronomy and Astrophysics, Committee on Planetary and Lunar Exploration, Committee on Solar and Space Physics, Committee on Earth Studies, Committee on International Space Programs) will provide input based on discussions at their regularly scheduled meetings over the period. The Space Studies Board will pay explicit attention to ensuring that participation from its committees draws on a mix of individuals who have knowledge of and experience with large, medium, and small space missions. The draft report will go to external review in December 1999 and will be released in February 2000.

---

[2]Space Studies Board, National Research Council, *An Integrated Strategy for the Planetary Sciences: 1995-2010*, 1994; SSB, *The Role of Small Missions in Planetary and Lunar Exploration*, 1995; SSB, *The Role of Small Satellites in NASA and NOAA Earth Observation Programs*, 2000; SSB, *A New Science Strategy for Space Astronomy and Astrophysics*, 1997; SSB, *A Science Strategy for Space Physics*, 1995; SSB, *Assessment of Recent Changes in the Explorer Program*, 1996; SSB, *Scientific Assessment of NASA's SMEX-MIDEX Space Physics Mission Selections*, 1997; SSB and Aeronautics and Space Engineering Board, *Reducing the Cost of Space Science Research Missions*, 1997.

C

Information Sought from Space Studies Board Discipline Committees

To obtain the information needed to carry out the three tasks assigned to it (Appendix B), the Ad Hoc Committee on the Assessment of Mission Size Trade-offs for Earth and Space Science Missions asked four of the Space Studies Board's discipline committees (the Committee on Astronomy and Astrophysics, the Committee on Earth Studies, the Committee on Planetary and Lunar Exploration, and the Committee on Solar and Space Physics) and one of the interdisciplinary committees (the Committee on International Space Programs) to answer the following questions:

1. Are there arguments for having a spectrum of mission sizes to achieve near-term (10 years) and far-term (10 to 20 years) goals in your discipline?[1] What are these arguments? Please cite relevant SSB strategy reports or other reports where appropriate. Relevant arguments might include laws of physics, cost and budget, timeliness and time-to-flight, institutional opportunities to participate, technology and technology readiness, and risk tolerance.

2. What are examples of existing, planned, or proposed missions along that spectrum of mission sizes? Examples should relate to missions that address high-priority science goals.

3. What criteria would you develop for evaluating the mix of missions you have chosen as examples?
   - Is the mix affected by the availability of off-the-shelf hardware?
   - Has the recent emphasis on various low-cost missions influenced your perceptions of what the best mix should be?
   - Has the recent support for low-cost missions improved or weakened near-term science in your discipline?
   - What is the impact of international cooperation on the mix of missions you chose? Criteria for evaluating the mission mix might include resilience, robustness, and satisfactory rates of progress against established science goals.

4. In applying these criteria to NASA's portfolio of missions in the NASA strategic plan for your discipline, what do you observe? To what extent do the projects planned or sponsored by other national space agencies

---

[1] For the purposes of this study, NASA defined small missions as those costing less than $150 million, medium-size missions as between $150 million and $350 million, and large missions as more than $350 million.

address discipline strategic goals not currently addressed in NASA's strategic plans? Where do such projects fit in the spectrum of missions defined below?

The discipline committees were asked to provide written responses to these questions as input to the ad hoc committee's information gathering and subsequent deliberations.

D

Meeting Agenda

WEDNESDAY, SEPTEMBER 8, 1999

8:00 am	*Breakfast—Woods Hole Study Center*
8:30	Chair's Remarks
9:00	Joint Session with Executive Committee of the Space Studies Board
Noon	*Lunch*
1:00 pm	Committee Meets Independently
	• Continue discussing task, approach, committee inputs
	• Produce report outline and writing assignments
5:30	*Adjourn*
5:45	*Reception and Clambake*

THURSDAY, SEPTEMBER 9, 1999

7:30 am	*Breakfast–Woods Hole Study Center*
8:30	Chair's Remarks
9:00	Committee Discussion and Work Session
	• Begin organizing material and drafting sections of report
Noon	*Lunch*
1:00 pm	Resume Work Session
3:30	Committee Discussion (Prepare for Joint Session with Executive Committee of the Space Studies Board)
	• Discuss initial drafts
	• Identify information and data gaps
	• Identify follow-up questions and requests from discipline committees
	• Set deadlines for next draft
	• Review report schedule and any other action items
4:30	Teleconference with Steering Committee Member George Paulikas
5:30	*Adjourn*

FRIDAY, SEPTEMBER 10, 1999

9:00 am	Joint Session with Executive Committee of the Space Studies Board
	• Discuss outline, approach, and organization of report
	• Discuss preliminary set of key issues/recommendations to be made
	• Discuss deadlines set for meeting report schedule
11:00	Wrap-Up
Noon	*Lunch and Adjourn*

E

Material Provided by Space Studies Board Discipline Committees

Two of the assignments for the Space Studies Board discipline committees (see Appendix C) were to identify examples of ongoing and planned missions in different sizes that address near-term (10 years) and far-term (10 to 20 years) science objectives in the discipline, and to identify criteria for evaluating the mix of missions chosen for the portfolio. This appendix presents the material submitted by the board's discipline committees in response to the assignment. In addition, each discipline summary includes a table showing a *nonexhaustive* set of ongoing and planned missions (including international missions) and their scientific objectives or parameters, life-cycle costs, size (as defined by NASA for the study), status, and estimated lifetimes.

EARTH SCIENCE

Arguments for a Portfolio of Mission Sizes

There are powerful arguments for having a broad portfolio of mission sizes to achieve near-term (10 years) and far-term (10 to 20 years) goals in Earth science. In each application, the scientific objectives and the specific mission goals will generally dictate the size range of the spacecraft. The need continues for the capabilities offered by several larger platforms such as NASA's Earth Orbiting System Chemistry Satellite (EOS-CHEM) mission: multiple instruments, extensive redundancy, and long lifetimes. Smaller missions aimed at well-focused and relatively short-term objectives allow for separate new starts to maintain pace with emerging scientific interests and needs. NASA's Earth System Science Pathfinder (ESSP) program is an excellent example of this small mission paradigm.[1] Smaller spacecraft can also reduce the time to "first science," if the instruments are already available (i.e., no development is required).

---

[1] The ESSP program was initiated by NASA in 1996. The program is intended to apply the spirit of FBC to the Earth Science Enterprise. Proposals submitted by the community in response to announcements of opportunity must offer principal-investigator-led, end-to-end flight missions within tight constraints. The proposed missions must also have (1) a significant and well-focused science objective, (2) a cap on the cost to NASA, typically $120 million, (3) minimal reliance on unproven or high-risk technology, (4) time to launch of less than 3 years, and (5) flight mission duration of nominally 2 years. NASA's plan is to conduct ESSP opportunities every 2 years or so, selecting 2 missions each round.

Larger Spacecraft

The principal arguments in support of larger spacecraft are (1) beneficial aggregation of instruments and (2) accommodation adequate to support large instruments. Increased science requirements or the need for more comprehensive data sets lead to significant growth in several instruments central to Earth science applications.[2] This growth is driven by increases—sometimes by large factors—in the required spatial resolution, number of spectral bands, and signal amplitude resolution (bits per sample).[3] These and similar enhancements call for larger apertures, higher data rates, greater power load, more severe thermal control requirements, and more stringent pointing requirements, among other things. The demand from the scientific community for increased performance has outstripped the ability of technology to keep pace.[4]

Smaller Spacecraft

The principal arguments in support of smaller spacecraft are that (1) they suffice to meet modest, well-focused objectives, (2) they minimize the time to science, and (3) they reduce the cost of achieving certain science objectives. A significant corollary is that with more new programs, there is a greater potential to increase the number of investigators and students involved from start to finish in space-based Earth observation. Small satellites, and by implication less complex missions, tend to be most effective when directed toward well-focused, short-term objectives. They can provide a rapid response for some missions if the required instruments have already been developed, presumably through a separate, adequately funded preparatory program. The concept of small satellites may also, however, connote higher risk and development shortcuts.[5] "Cheaper" should be interpreted as "more cost-effective" rather than "short-changed" if the faster-better-cheaper (FBC) philosophy is to have value in the long run. Small satellites also have the reputation of being short-lived, although there is no intrinsic reason for this to be so.[6] New technology coupled with cost-effective design and implementation practices would make multiple small satellite options appealing alternatives to large, multi-instrument systems for certain science applications. However, these faster-better-cheaper and (sometimes) smaller missions must not be invented ad hoc but should fall within a comprehensive long-term science plan.

---

[2]Space Studies Board, National Research Council, *Earth Observations from Space: History, Promise, Reality*, National Academy Press, Washington, D.C., 1995.

[3]As noted in Chapter 3, research advances in long-term weather forecasting and the development of climate computer models have increased the requirements for horizontal and vertical resolution for temperature and moisture atmospheric profiles. The need for 1-km vertical resolution will necessitate larger instruments.

[4]SSB, *Earth Observations from Space*, 1995.

[5]See Space Studies Board, National Research Council, *The Role of Small Satellites in NASA and NOAA Earth Observation Programs*, National Academy Press, Washington, D.C., 2000, p. 54, which discusses the risks involved in employing small satellites:

The small satellite approach carries with it several risks concerning scientific return:

1. Rapid development missions are often focused on "small" problems. Missions are not designed for long life and are sometimes viewed as "one shot" opportunities.
2. Missions employing small satellites are more likely to be developed as part of a program of technology demonstrations as opposed to a program in which the science return is paramount.
3. Small missions require a well-defined focus to keep them simple and the cost low. This approach may not work well for scientific studies that require measurements of many processes.
4. Data processing and distribution may be related to relatively lower priority, thus making it difficult for nonproject scientists to gain access to the data. This problem could be exacerbated in the case of missions led by a principal investigator (PI) should research investigations become centered on an individual's personal scientific interests.
5. With more single-sensor missions, the proportion of funds spent on satellite hardware and launch vehicles will increase. Such funds might otherwise be spent on scientific research.

[6]For example, one of the early and very small Transit navigation satellites maintained full operational status for 22 years.

A satellite should be as large as necessary, but no larger than needed to carry out its mission; the overall program should be funded sufficiently, but no more than necessary to complete its tasks. Objective cost trade-offs can be used to determine the right size for a spacecraft.[7]

If budget is the dominant consideration, low-cost missions generally bear higher risk, both in development and in operation, as was made all too clear by Lewis and Clark and the more recent Mars missions.[8] If reliability is compromised by a lack of redundancy or for other cost-reduction reasons, the spacecraft life may be curtailed. Some of the smaller, cheaper programs have been successful (e.g., QuikSCAT), but in general, short schedules and mission success depend on the availability of previously developed instruments and a baseline spacecraft as well as efficient and effective mission management and operations.[9]

Need for Long-Term Measurements

Programs such as the Earth Observing System (EOS) that have long-term science objectives (the measurement of climate variables, for example) cannot be replaced by short-term limited missions. Long-term and broad-based systematic measurements require a long-term commitment. Such a commitment could be met by a mix of satellite sizes if well-planned and coordinated.[10] The SSB's Committee on Earth Studies recently completed an analysis of issues related to the transitioning of NASA research satellite instrumentation for NOAA's operational use and conducted a workshop on joining the Integrated Program Office (IPO)/NPOESS and NASA/Earth Science Enterprise capabilities for climate research. Among the conclusions, the Committee on Earth Studies notes that climate studies require long-term measurements, revision and independent validation of the algorithms, ability to reprocess older data, and good characterization and calibration of instruments.[11]

---

[7]SSB, *Earth Observations from Space*, 1995, p 134.

[8]The Small Spacecraft Technology Initiative (SSTI) was developed by NASA's Office of Space Access and Technology to advance the state of technology and reduce the costs associated with the design, integration, launch, and operation of small satellites. In July 1994, TRW and CTA Space Systems were each awarded a contract by NASA to design and launch small Earth-observing satellites, one named "Lewis," the other named "Clark." Both contracts called for substantial infusion of new technology into both payload and spacecraft bus and for delivery of the satellites to launch in only 24 months following contract start. Both missions were unsuccessful. In the case of Lewis, the satellite development was completed within the allotted 24-month period and, after a 1-year delay before its Athena-I (LMLV-I) launch vehicle was deemed flight-ready, the satellite was successfully placed into its initial orbit in August 1997 and was subsequently lost. The Clark mission suffered excessive schedule delays and projected cost growth, ultimately leading to termination of the contract.

[9]"The often complex evaluation of whether the use of a small satellite is appropriate is driven by mission-specific requirements, including those related to the policy and execution of the program, fiscal constraints, and the scientific needs of the end users. Considering the many issues involved, the design of an overall mission architecture, whether for operational or research needs, requires a complete risk-benefit assessment for each particular mission. For some missions, a mixed fleet of small and large satellites may provide the most flexibility and robustness, but the exact nature of this mix will depend on mission requirements" (SSB, *The Role of Small Satellites*, 2000, p. 5).

[10]An excellent example is provided by ongoing measurements of atmospheric ozone, particularly over Antarctica. TOMS, the Total Ozone Mapping Spectrometer, has flown on four different spacecraft since 1978. TOMS was one of several instruments on three previous large satellites, but the current mission is a small (295 kg) Earth probe. Since TOMS is only a 35-kg instrument, that option works well. From a climate point of view, however, ozone is only part of the story. Climate, or the trend in long-term weather, depends on atmospheric gases and particulates, clouds, moisture, temperature, and their circulation and interaction with the surface, especially the sea's surface. Climate observations must track all of these key variables for many years. Several of the instruments required are considerably larger than can be accommodated on a small spacecraft.

[11]Issues raised by the integration of research and diverse climate measurements into the operational NPOESS have been studied by the SSB Committee on Earth Studies. Its phase I report, *Issues in the Integration of Research and Operational Satellite Systems for Climate Research: I. Science and Design*, emphasizes the potential science value of operational weather observations and examines the fit between NPOESS and the climate requirements with respect to a set of eight environmental variables. *Issues in the Integration of Research and Operational Satellite Systems for Climate Research: II. Implementation*, which is forthcoming, focuses on technology and related spacecraft issues.

Example of a Portfolio of Mission Sizes

Examples abound in the Earth sciences of a wide variety of Earth-observing missions and their satellites. Selected Earth observation missions drawn from NASA's current plan[12] are provided in Table E.1. Missions shown in the table are meant to be representative of the current and projected mix of satellite sizes.

Terra, the first satellite in the EOS program to be launched, and its companions, EOS-PM (now Aqua) and EOS-CHEM, are survivors after a decade of recurrent revisions of NASA's EOS plans. All three are large[13] satellites, with a design life of 5 or 6 years. The original EOS plans for these missions spanned 15 years, to be implemented by a series of three nearly identical satellites in each sequence. All were to share a large common bus, which promised certain economies of scale to reduce their aggregate cost; however, the common-bus approach has been abandoned. Landsat-7 has a long history, the Landsat series having been returned to NASA after a complicated path originating at NASA in the 1970s, passing first to the commercial world, next to NOAA, and then back to NASA. It is part of a sequence of Earth-imaging systems that provide an important source of data for environmental and land process studies. Whether Landsat-like observations will continue beyond Landsat-7 is under discussion.

The weather satellites GOES, POES, and NPOESS are first and foremost operational weather satellites. Their data, which now span more than 40 years, provide the most important space-based record of climate and environmental variables. These data are and will continue to be in demand for scientific studies. The NPOESS Preparatory Project satellite appears to be unique, providing a one-time opportunity to validate new technologies and instruments of interest for NASA research and NOAA's operational activities. UARS, launched on the space shuttle in 1991, continues to provide an important series of upper atmospheric observations. Its large size is as much a reflection of the relatively unconstrained mass limitations of the shuttle as it is a consequence of its mission. The Ocean Topography Experiment (TOPEX)/Poseidon and Jason-1 are the two currently funded precision ocean altimeter missions in which the United States has a major role. SeaWiFS, an instrument flown in a precedent-setting commercial partnership with the private sector, provides an important source of ocean color data.[14] EO-1, the first of the New Millennium satellites,[15] is designed primarily as a technology demonstration platform; science (if any) is secondary.

The Gravity Recovery and Climate Experiment (GRACE) (ESSP-2) is a good example of the faster-better-cheaper system encouraged by the ESSP program. Total mission costs are about twice those billed to NASA, thanks to major foreign contributions. GRACE would not be possible without extensive prior developments, including a large science base and an existing spacecraft design, Challenging Minisatellite Payload (CHAMP), for a European mission. Note also that the science objective of GRACE cannot be met by one satellite, as it takes two satellites to capture the level of detail in Earth's gravity field to be mapped by GRACE. In turn, a twin-satellite mission would not be feasible, technically or financially, without the motivation afforded by the spirit of FBC.

---

[12]Available at <http://www.earth.nasa.gov/missions/2002/index.html>.

[13]Small, medium, and large are meant to convey total system cost, as defined by NASA. Because such numbers are often controversial and the cost may be complicated by extensive international participation, spacecraft mass is also listed, where known, to provide an objective indication of the size of each satellite.

[14]NASA's coastal zone color scanner (CZCS), hosted on a Nimbus-7 satellite in the 1970s, first proved the importance of such data. Neither CZCS nor its equivalent was adopted by NOAA as an operational instrument.

[15]The New Millennium program is a technology demonstration program to validate technology in spaceflight that will lower the risks to future science missions using the technologies. The program draws on existing government-funded research and development efforts.

APPENDIX E

TABLE E.1 Selected Earth Science Missions

Spacecraft	Parameters/Goals	Mission Size/Mass/Life-Cycle Costs (real $)	Status	Time Scale of Observation
ONGOING				
QuikSCAT	Ocean winds, sea ice storm patterns	Small, 970 kg, $95 million	Ongoing	1999- (2-yr design)
SeaWiFS	Ocean color (commercial system)	N/A[a]	Ongoing	1997
TRMM	Tropical rainfall	Medium, 3,600 kg, $296 million	Ongoing	1996-
POES	Meteorology/weather (original TIROS–NASA)	Large, 2,250 kg	Several in orbit	1960-2010 (2-yr design)
GOES	Meteorology/weather (original SMS–NASA)	Large, 2,100 kg	Several in orbit	1974-
UARS	Upper atmosphere (10 instruments)	Large, 6,800 kg, $750 million	Ongoing	1991-
TOPEX/Poseidon	Ocean altimetry	Large, 2,500 kg, $450 million	Ongoing	1992- (5-yr design)
Landsat-7	Medium-resolution, multiband Earth imager	Large, 2,200 kg, $447 million	Ongoing	1999- (5-yr design)
Terra (EOS-AM)	Earth and atmospheric study	Large, 5,190 kg, $1.2 billion	Ongoing	1999- (5-yr design)
PLANNED				
ESSP-2 (GRACE)	Gravity field mapping (two spacecraft constellation) and cooperation with Germany	Small, NASA, $86 million; Germany, $45 million	Planned	2001 (2-yr design)
EO-1	Technology: land mapping instruments	Medium, 530 kg, $162 million	Planned	2000
ICESat	Laser altimetry measuring ice-sheet topography and temporal changes, cloud and atmospheric	Medium, $227 million	Planned	July 2001
Jason-1	Ocean altimetry	Medium, 500 kg, $160 million (French-led mission; NASA contribution, $94 million)	Planned	2001 (5-yr design)
Aqua (EOS-PM)	Earth and atmospheric study	Large, 3,120 kg, $880 million	Planned	2001 (6-yr design)

continued

TABLE E.1 Continued

Spacecraft	Parameters/Goals	Mission Size/Mass/Life-Cycle Costs (real $)	Status	Time Scale of Observation
EOS-CHEM	Tropospheric chemistry	Large, 2,970 kg, $670 million	Planned	2002 (5-yr design)
NPP	Flight of opportunity, plus new instruments	Large, Estimated $800 million: $500 million from NASA; remainder from DOD and NOAA	Planned	2005 (5-yr design)
NPOESS	Meteorology/weather	Large, $ N/A[a]	Planning	2008-2020 (10-yr design)

[a]NA, not available.

Criteria for Evaluating the Mission Mix

Previous Space Studies Board reports identified several criteria that can be used for evaluating the balance of missions in the Earth science portfolio.[16] In addition to being assessed on their individual scientific merit, missions are assessed based on the extent to which they do the following:

- Address the high-priority scientific goals;[17]
- Serve the needs of the U.S. Global Change Research Program;
- Reflect the need for sustained, long-term observations as well as for short-term, focused research;
- Respond to the need to facilitate the transition from research to operations;
- Incorporate and justify appropriate new technology without being technology-driven;
- Involve the science community in overall planning;
- Allow for increased risk in individual, specialized missions;
- Balance spacecraft design against instrument reliability and mission time horizon; and
- Recognize the need for stable funding for a diligent, sustainable, and productive program.

Several of the criteria underscore the need for long-term planning in the space-based Earth science program, as stated in two Space Studies Board reports:

> The planning process should be an orderly one that is aimed at minimizing the continual changes. The process also must include more attention to intricate issues associated with . . . ensuring the continuity of key long time-series measurements.[18]

and

---

[16]Space Studies Board, National Research Council, "Report of the Task Group on Assessment of NASA's Plans for Post-2002 Earth Observing Missions," letter to Ghassem Asrar, associate administrator, Office of Earth Science, NASA, April 8, 1999, pp. 5-9.

[17]Board on Sustainable Development, National Research Council, *Global Environmental Change: Research Pathways for the Next Decade*, National Academy Press, Washington, D.C., 1999.

[18]Space Studies Board, National Research Council, "Report of the Task Group on Assessment of NASA's Plans for Post-2002 Earth Observing Missions," letter to Ghassem Asrar, associate administrator, Office of Earth Science, NASA, April 8, 1999, p. 15.

APPENDIX E   77

... the CES [Committee on Earth Studies] believes that the needs of research in the Earth sciences and applications should not be continually deferred until the development of new, unproven technologies. The prospect of lower cost is always attractive, but the practice of placing new technology developments ahead of the conduct of basic and applied research has been disruptive to the Earth sciences for more than two decades.[19]

PLANETARY SCIENCES

Arguments for a Portfolio of Mission Sizes

The Space Studies Board's strategy report for the planetary sciences, *An Integrated Strategy for the Planetary Sciences: 1995-2010*, covers a diversity of topics and objectives, including studies of protoplanetary disks, planetary systems about other stars, primitive bodies, the origin and evolution of life, the surfaces and interiors of solid bodies, and planetary atmospheres, rings, and magnetospheres.[20] The scientific, technical, and operational aspects of planetary exploration require a range of mission sizes.[21,22] Large missions are necessary to approach future high-priority scientific goals such as a sample return from a comet nucleus or from the surface of a planet or satellite, a comprehensive survey of a giant planet (with atmospheric and satellite probes), and extensive exploration of Mars in preparation for human missions.[23] Small missions are critical for continuing the introduction and infusion of new technology and for addressing very tightly focused scientific goals. The current mix of planetary-science missions seems to reflect this balance. However, there is a growing emphasis on medium-size missions that is affecting the balance.

Example of a Portfolio of Mission Sizes

Table E.2 shows examples of planetary science missions that span the range of sizes. It includes planned missions as well as missions that are currently operating. Many of the ongoing and planned missions are medium-size missions (or nearly so) in the Discovery line. The few low-cost missions are either at the low end of Discovery (e.g., Contour) or are non-NASA or international missions for which the total costs are uncertain (e.g., Nozomi). Some missions (e.g., Europa Orbiter and Pluto-Kuiper Express) have officially been classified as large missions, but they are near the medium-size cap of $350 million. Many larger missions are either conglomerate missions (Mars Sample Return) or missions involving international collaboration (e.g., Rosetta or Mars Express), whose costs are uncertain.

---

[19]SSB, *Earth Observations from Space*, 1995, p. 134.

[20]Space Studies Board, National Research Council, *An Integrated Strategy for the Planetary Sciences: 1995-2010*, National Academy Press, Washington, D.C., 1994, pp. 3-6.

[21]"... Priority scientific investigations [identified by the *Integrated Strategy*] can be addressed by the full gamut of techniques, including small (inexpensive) and large (expensive) robotic probes, ground- and space-based observatories, and laboratory studies and theoretical modeling" (SSB, *Integrated Strategy*, p. 186).

[22]"Many diverse objects across the solar system must be studied to achieve the broad goals of planetary and lunar exploration [outlined by COMPLEX]. An effective program for lunar and planetary exploration also dictates a mix of mission sizes, ranging from comprehensive missions with multiple objectives, such as Galileo and Cassini, down to relatively low cost missions, such as those in the Discovery program" (SSB, *The Role of Small Missions*, 1995, p. 27).

[23]"The long travel times between Earth and the outer solar system require long-lived components, specialized power systems, and complex, high-powered communications. This implies that, with current technology, any mission sent beyond the asteroid belt must be very capable. In addition, many [priority studies] require concurrent coordinated observations between the different components of a particular planet or comet (e.g., simultaneous in situ and remote-sensing observations of Titan's atmosphere by Huygens and Cassini, respectively). Thus, COMPLEX believes that many solar system missions, especially those to the outer solar system, cannot be adequately accomplished by reconfiguration of large spacecraft into one or more small spacecraft" (SSB, *Integrated Strategy*, pp. 182-183).

TABLE E.2 Selected Planetary Exploration Missions

Spacecraft	Parameters/Goals	Mission Size/Mass/ Life-Cycle Costs (real $)	Status	Time Scale of Observation
ONGOING				
NEAR	Asteroid rendezvous	Medium, 503 kg, $224 million	Ongoing	1996 4-yr cruise 1-yr operations
Mars Global Surveyor	Mars mapping mission	Medium, 1,030 kg, $273 million	Ongoing	1996 5 years
Nozomi	Atmosphere and ionosphere of Mars; U.S. contributed NMS instrument	Medium, Japanese-led mission; U.S. contribution, $6 million	Ongoing	Launched 1998
Stardust	Collect comet material	Medium, 380 kg, $205 million	Ongoing	February 1999 7 years
Galileo	Jupiter orbiter and probe	Large, $1,425 million	Ongoing	Launched 1988, in orbit 1995-
Cassini	Saturn system including Titan probe	Large, 5,650 kg, $2,550 million	Ongoing	October 1997, over various timescales up to 11 years
PLANNED				
Contour	Imaging and spectral maps of three comets	Small, 489 kg, $144 million	Planned	2002 6 years, 3 flybys
Genesis	Solar wind sample return	Medium, 648 kg, $216 million	Planned	January 2001 2-yr operations
Messenger	Mercury Orbiter	Medium (Discovery), $339 million	Planned	Launch 2004 1-yr orbital operations
Deep Impact	Image subsurface of a comet	Medium, 600 kg, $240 million	Planned	Launch January 2004 1.5 years
Mars 2001 (orbiter and lander)	Mars geochemical mapper	Large, 1,460 kg, $415 million	Planned	2001, 3-yr orbiter
Europa Orbiter	Europa search for oceans	Large, 1,600 kg, $460 million	Planned	Launch in 2003, 6 years
Mars Sample Return (1 French orbiter, 2 U.S. landers)	Mars sample return	Large, 1,800 kg each, $1,100 million	Planned	2005 and 2007 launches, sample return in 2010
Rosetta (ESA mission, NASA providing 4 instruments)	Comet lander and surface investigation	Large total (U.S. contribution, $39 million)	Planned	January 2003 10-yr cruise plus landed operations

continued

TABLE E.2 Continued

Spacecraft	Parameters/Goals	Mission Size/Mass/ Life-Cycle Costs (real $)	Status	Time Scale of Observation
Mars Express (ESA mission; NASA providing components of ASPERA-3 (Energetic Neutral Atoms Analyzer) plus other: radio frequency section, transmitter, antenna subsystems for the radar instruments)	Interaction of solar wind with Mars atmosphere	Large total (U.S. contribution, $6.6 million for ASPERA, $27 million for other contributions)	Planned	June 2003 3 years
Pluto-Kuiper Express	Surface and atmosphere	Large, 225 kg, $354 million	Planned	2004 More than 9 years

Criteria for Evaluating the Mission Mix

The following criteria are proposed for evaluating the current mix of missions for solar system exploration:

- Addresses high-priority scientific goals;
- Optimizes science return for the money spent;
- Exhibits compatibility between mission goals and scale;
- Demonstrates a balanced-risk strategy;
- Considers future application of new technologies;
- Shows balance between technology and science;
- Involves community in mission/instrument/technology selection;
- Promotes stable funding and continuous planetary exploration;
- Is consonant with Deep Space Network (DSN) and mission operations and data analysis (MO&DA) support;
- Uses diverse modes of mission implementation (principal-investigator-led, university-industry-NASA team, NASA-led); and
- Incorporates education and public outreach.

SPACE AND SOLAR PHYSICS

Arguments for a Portfolio of Mission Sizes

Solar and space physics are mature sciences in which exploration, discovery, and observations have been carried out in space for more than four decades. The disciplines are now essentially beyond the exploration and discovery phases, although—remarkably—discoveries are still made. The disciplinary maturity requires that future missions must address sophisticated questions that require a coordinated and novel approach—typically involving multiple instruments and even multiple spacecraft—to measure the many physical variables involved and to separate spatial from temporal physical effects. In the past, the disciplines were driven by science questions and priorities and not by mission size per se. That approach—science-driven missions—led naturally to a portfolio of mission sizes that included small, medium, and large missions. As explained below, the diversity and maturity of Sun-Earth Connection (SEC) science and current scientific priorities, even though they are being implemented using FBC principles, continue to necessitate medium-size and occasionally large missions. Thus, the new SEC Roadmap sets forth a community science plan that requires Explorers (small missions), Solar-Terrestrial Probes (medium-size missions), and Frontier Probes (medium and large missions) for implementation. Interdisciplinary missions to the planets (space physics and planetary science) would most likely require large missions (more than $350 million) unless conducted through the Discovery program.

Example of a Portfolio of Mission Sizes

Table E.3 includes an array of currently operating and planned missions in space and solar physics. Some of the missions are important both for addressing science questions and for understanding current space weather conditions.

TABLE E.3 Selected Space Physics Missions

Spacecraft	Parameters/Goals	Mission Size (Program)/ Mass/Life-Cycle Costs (real $)	Status	Time Scale of Observation
ONGOING				
IMP-8	Near-Earth solar wind monitor	Small (Explorer), $ n/a	Ongoing	Launched 10/25/73; far into extended-phase operations
SAMPEX	Observations of solar energetic particles, cosmic rays, precipitating relativistic electrons	Small (SMEX), $80 million	Ongoing	Launched 7/3/92; continues to 2003
FAST	Electron and ion acceleration, plasma dynamics above auroral zone	Small (SMEX), $74 million	Ongoing	Launched 8/21/96, 1-year primary plus extended mission as permitted
SNOE	Nitric oxide density and variation in Earth's upper atmosphere	Small (UNEX), $12 million	Ongoing	Launched 2/26/98
TRACE	Solar magnetic structures, heating of solar atmosphere, flare onsets	Small (SMEX), $72 million	Ongoing	Launched 4/1/98, 1-year primary phase, in extended phase
Geotail	Magnetotail dynamics, the near-Earth neutral line, and the magnetopause	Medium, $150 million (Japanese-led mission)	Ongoing	Launched 7/24/92, 2-year primary and extended phase through 2002
ACE	Interplanetary particles, composition, energy spectrum, solar wind plasma	Medium (Explorer), $215 million	Ongoing	Launched 8/25/97, 3-year primary plus 2-year extended mission
Wind	Comprehensive measurements of solar wind plasmas, fields, and radio waves upstream of Earth	Large, $360 million	Ongoing	Launched 11/1/94, extended to 2003
SOHO	Measurements of solar electromagnetic emissions, the solar interior, the inner heliosphere, and the solar wind (ESA-led mission)	Large, $430 million, cost is U.S. portion only for instruments, launch, and tracking and missions support	Ongoing	Launched 12/2/95, now in extended phase through 2005

continued

TABLE E.3 Continued

Spacecraft	Parameters/Goals	Mission Size (Program)/ Mass/Life-Cycle Costs (real $)	Status	Time Scale of Observation
Polar	Energy flow into magnetosphere and the ionosphere in polar regions	Large, $420 million	Ongoing	Launched 2/25/96, extended to 2003
PLANNED				
HESSI	Solar flare high-resolution spectroscopy and imaging (3 keV to 2 MeV)	Small (SMEX), 281 kg, $76 million	In development	Launch 7/4/00, planned 3-year lifetime
IMEX	Dynamics of inner magnetosphere and storms	Small (UNEX), 350 lb, $13 million	Under study	Launch 8/00
IMAGE	Imaging magnetospheric plasma, boundary layers, and auroras	Medium (MIDEX), 536 kg, $154 million	Waiting for launch	Launch 3/15/00, 2-yr primary plus extended phase
TIMED	Energy flow and dynamics in the 60- to 180-km region of Earth's atmosphere, by remote sensing	Medium (STP), $208 million	In development	Launch 7/01, 2 years of operation
Solar B	Solar magnetic field evolution at photosphere, lower corona	Medium, 875 kg, $154 million (Japanese-led mission)	Phase A/B	Launch 2003
STEREO	Observe stereoscopically CMEs and solar energetic particles from the photosphere to Earth	Medium (STP), $318 million (2 spacecraft)	Phase A/B	Launch 2004 on a Taurus
Magnetospheric Multiscale	Turbulence, reconnecting plasma entry at plasma boundaries	Medium (STP), $ N/A^a	STD in process	Launch 2005 on a Delta 7325
Global Electrodynamic Constellation	Plasma and electrodynamic coupling in Earth's upper atmosphere/ ionosphere	Medium (STP), $ N/A	STD in process	Launch 2007 on a Delta 7325
Magnetotail Constellation	3-D dynamic imaging of the outer magnetosphere	Medium (STP), $ N/A	STD in process	Launch 2008 on a Delta 7325
Solar Polar Imager	Solar polar fields, origin of solar wind and activity cycle	Medium (Frontier Probe), $ N/A	Under study	Launch 2010
Solar Probe	Coronal heating, acceleration mechanism of solar wind	Medium (Outer Planets), $156 million	Planning, under study	Launch 2010 (?)
Interstellar Probe	Explore outermost heliosphere and interaction with local instellar medium	Large (Frontier Probe), $ N/A	Under study	Launch ~2010
CLUSTER II	3-D study of plasma using 4 identical spacecraft at magnetospheric boundaries and magnetotail, separating space and time variations	Large (ESA/NASA mission, NASA contribution $14.3 million to replace the U.S. instruments)	In development	Launch June and July 2000 (by 2 Soyuz rockets)

^{a}N/A, not available.

Criteria for Evaluating the Mission Mix

The recent Roadmap exercises conducted by the Sun-Earth Connection theme reflect the portfolio of mission sizes recommended by the solar and space physics science communities to address current scientific priorities.[24] Determined and pragmatic efforts were made in these exercises to adhere to NASA's expectations of FBC and NASA's budgetary trends, with the result that most of the missions recommended were made to fit into the Solar-Terrestrial Probe (STP) program. An STP is capped at $250 million and has an anticipated launch rate of about one every 18 months. Quoting from the 1997 Roadmap document:

> The majority of the candidate missions described in the Roadmap would be implemented under NASA's STP program, which offers a continuous sequence of flexible, cost-capped missions designed for the systematic study of the Sun-Earth system. The strategy embodied in the STP mission line is to use a creative blend of in-situ and remote sensing techniques and observations, often from multiple platforms, (i) to provide understanding of solar variability on time scales from a fraction of a second to many centuries, with an underlying activity cycle of approximately 11 years; and (ii) to determine cause (solar variability) and effect (planetary and heliospheric response) relationships over vast spatial scales. The latter objective generally requires innovative multi-spacecraft and/or missions operating concurrently.[25]

A portfolio of missions to carry out the scientific objectives in space and solar physics is reflected in Table E.3, which shows that both ongoing and planned missions span the full spectrum of small, medium, and large sizes. Small and medium missions in the plan have focused scientific objectives. Certain other scientific requirements can be met only by large missions: a long observation time line (solar variations or sunspot cycle effects); multiple instruments of high resolution (microphysics of particles and fields); highly stable platforms (for remote observations); vast physical parameter ranges in the operating environment (heliospheric observations outward to interstellar space); interdisciplinary missions (planetary missions to investigate both the planet and its environment); and use of multiple spacecraft or constellations (to separate spatial and temporal effects or to make complementary observations simultaneously).

ASTRONOMY AND ASTROPHYSICS

Arguments for a Portfolio of Mission Sizes

The importance of having a mix of mission sizes and costs to pursue space astronomy and astrophysics has long been recognized by the astronomy and astrophysics community. For example, the previous astronomy and astrophysics survey committee (the Bahcall committee) noted in its 1991 report, *The Decade of Discovery in Astronomy and Astrophysics*, that a vigorous program in space astronomy and astrophysics requires a proper mix between small, moderate, and large missions.[26]

The goals identified by the Bahcall committee and NRC strategies[27] are broad and diverse and can be answered in a cost-effective way only through a coordinated program. In addition to suiting the scientific and physical requirements, a mix of mission sizes provides for continuity and follow-up in the various subfields of space astronomy and astrophysics. To achieve that diversity, the Bahcall committee recommended that the Explorer program be substantially expanded to allow flying six Delta-class astronomy and astrophysics Explorer missions and five SMEX-class missions for astrophysics during the 1990s.[28] Most of the Delta-class missions would fall into the medium cost category.

---

[24]NASA, Office of Space Science, *Sun-Earth Connection Roadmap: Strategic Planning for 2000-2025*, 1999.

[25]NASA, *Sun-Earth Connection Roadmap: Strategic Planning for the Years 2000-2020*, 1997.

[26]National Research Council, Astronomy and Astrophysics Survey Committee, *The Decade of Discovery in Astronomy and Astrophysics*, National Academy Press, Washington, D.C., 1991, pp. 15-16.

[27]See, for example, Space Studies Board, National Research Council, *A New Science Strategy for Astronomy and Astrophysics*, National Academy Press, Washington, D.C., 1997.

[28]National Research Council, *The Decade of Discovery*, 1991, p. 23.

APPENDIX E   83

Example of a Portfolio of Mission Sizes

Table E.4 lists current and planned missions in astronomy and astrophysics. As shown, there are several small and medium missions under development; few are currently operating.

TABLE E.4 Selected Astronomy and Astrophysics Missions

Spacecraft	Parameters/Goals	Mission Size/Mass/ Life-Cycle Costs (real $)	Status	Time Scale of Observation
ONGOING				
SWAS	Submillimeter spectrum molecules in star-forming regions	Small, $97 million	Ongoing	December 1998-present
ACE	Particles, isotopic, elemental composition of planetary and interstellar space	Medium, $203 million	Ongoing	August 1997 2-5 years
FUSE	Far-UV spectrum, deuterium, H_2, hot gas	Medium, $204 million	Ongoing	June 1999 3 years
HST	Optical, UV, and near-IR observations	Large, $9.1 billion (including operations, data analysis, and use of shuttle)	Ongoing	April 1990-present
Compton Gamma Ray Observatory	Gamma-ray astrophysics	Large, $ N/A[a]	Ongoing	April 1991-present
Chandra X-ray Observatory	X rays, supernovae, compact stars, AGNs	Large, $2,800 million	Ongoing	July 1999 5+ year lifetime
XMM European-led mission with U.S. guest observer program	X rays to faint flux limits	Large, $ N/A	Ongoing	December 1999
PLANNED, Short-Term				
HETE-2	Gamma-ray bursts/fast response	Small, 125 kg, $23 million	Planned	2000 5-2 years
MAP	CMBR anisotropy <1 deg	Small, 800 kg, $149 million	Planned	Fall 2000 2 years
GALEX	UV surveys/galaxy evolution	Small, 280 kg, $76 million	Planned	2 years
CATSCAT	Origin and nature of gamma-ray bursts	Small (UNEX), $ N/A	Planned	July 2001
CHIPS	EUV spectrum, hot local ISM	Small, $12 million	Planned	April 2002 1 year

continued

TABLE E.4 Continued

Spacecraft	Parameters/Goals	Mission Size/Mass/ Life-Cycle Costs (real $)	Status	Time Scale of Observation
Swift	Gamma-ray bursts	Medium, 1,270 kg, $163 million	Planned	2003 3 years
FAME	All-sky stellar astrometry	Medium, 1,030 kg, $162 million	Planned	2004 5 years
GLAST	Gamma rays/AGNs, bursts, pulsars, supernovae remnants	Medium, 4,500 kg, $330 million	Planned	September 2005 5 years
SOFIA (aircraft)	Suborbital infrared/star, planet formation	Large, $1,351 million (including 20 years of operations)	Planned	November 2002 20 years
GP-B	Gyroscopes/test general relativity	Large, 3,300 kg, $556 million	Planned	Fall 2001
SIRTF	Infrared/brown dwarfs, protoplanetary disks, AGNS, distant galaxies	Large, 905 kg, $880 million	Planned	December 2001 2.5-5 years
ACCESS	Cosmic ray experiment	Large, $ N/A	Planned	2005
PLANNED, Long-Term				
SIM	Optical interferometer/parallax, proper motion, planet detection	Large, 5,000 kg, $900 million est.	Planned	June 2006 5 years
NGST	Near-IR/high-redshift galaxies	Large, 3,300 kg, $1,700 million	Planned	2008 5 years
LISA	Interferometer/gravitational radiation with contributions from several countries	Large, $ N/A	Under study	2009 6 years
Constellation-X	X rays (imaging and spectroscopy)	Large, $ N/A	Planned	2010 3-5 years
TPF	IR interferometer/planet detection, processes related to star and planet formation, AGNs	Large, $ N/A	Planned	2011 >5 years
FIRST/Planck	Image the anisotropies of the cosmic background radiation field over the whole sky	Large, $ N/A (ESA missions with NASA contributions)	Planned	2003 for FIRST, 2007 for Planck

[a]N/A, not available

Criteria for Evaluating the Mission Mix

Mission selection criteria were well described in a 1986 report by the NASA Advisory Council[29] and are excerpted below:

- Scientific merit
    —Significance of the scientific objectives
    —Potential for new discoveries and understanding
    —Generality of interest;
- Programmatic considerations
    —Feasibility and readiness
    —Infrastructure requirements
    —Cost effectiveness
    —Institutional implications; and
- Societal and other implications
    —Potential for stimulating technological development
    —Contributions to scientific awareness of the public
    —Contributions to international understanding
    —Contributions to national pride and prestige.

To evaluate the mix of mission sizes, the criteria need to be weighted according to the mission costs. For example, a large mission would need to have a very broad impact while a small mission might be selected to pursue a narrow science problem and to stimulate technological development. More risk could be accepted with medium-size and small missions.

---

[29]NASA Advisory Council, Space and Earth Science Advisory Committee, *The Crisis in Space and Earth Science: A Time for New Commitment*, 1986, pp. 55-58.

F

Acronyms and Abbreviations

ACCESS	Advanced Cosmic-ray Composition Experiment for the Space Station
ACE	Advanced Composition Explorer
AGN	active galactic nucleus
AIRS	Advanced Infrared Sounder
ALEXIS	Array of Low-Energy X-ray Imaging Sensors
AO	announcement of opportunity
ARISE	Advanced Radio Interferometry between Space and Earth
ASCA	Advanced Satellite for Cosmology and Astrophysics
AXAF	Advanced X-ray Astrophysics Facility (now Chandra X-ray Observatory)
CAA	Committee on Astronomy and Astrophysics
CATSAT	Cooperative Astrophysics and Technology Satellite
CES	Committee on Earth Studies
CFC	chlorofluorocarbon
CGRO	Compton Gamma Ray Observatory
CHAMP	Challenging Minisatellite Payload
CHIPS	Cosmic Hot Interstellar Plasma Spectrometer
CISP	Committee on International Space Programs
CMBR	cosmic microwave background radiation
CME	coronal mass ejection
COMPLEX	Committee on Planetary and Lunar Exploration
CRAF	Comet Rendezvous Asteroid Flyby
CSSP	Committee and Solar and Space Physics
CZCS	Coastal Zone Color Scanner
DMSP	Defense Meteorological Satellite Program
DOD	Department of Defense
DSN	Deep Space Network

EMI	electromagnetic interference
EO-1	Earth Orbiter-1
EOS	Earth Observing System
EOS-AM	Earth Observing System Morning Satellite (now Terra)
EOS-CHEM	Earth Observing System Chemistry Satellite
EOS-PM	Earth Observing System Afternoon Satellite (now Aqua)
ESA	European Space Agency
ESE	Earth Science Enterprise
ESSP	Earth System Science Pathfinder (program)
EUVE	Extreme Ultraviolet Explorer
FAME	Full-Sky Astrometric Mapping Explorer
FAST	Fast Auroral Snapshot Explorer
FBC	faster-better-cheaper
FIRST	Far Infrared and Submillimeter Telescope
FUSE	Far Ultraviolet Spectroscopic Explorer
GALEX	Galaxy Evolution Explorer
GCRP	Global Change Research Program
GEC	Global Electrodynamic Constellation
GGS	Global Geospace Science (program)
GLAST	Gamma Ray Large Area Space Telescope
GO	guest observer
GOES	geostationary operational environmental satellites
GP-B	Gravity Probe-B
GPS	Global Positioning System
GR	general relativity
GRACE	Gravity Recovery and Climate Experiment
GSFC	Goddard Space Flight Center
HESSI	High-Energy Solar Spectroscopic Imager
HETE	High-Energy Transfer Explorer
HIRS	High-Resolution Infrared Sounder
HST	Hubble Space Telescope
ICESat	Ice, Cloud, and land Elevation Satellite
IIP	Instrument Incubator Program
IMAGE	Imager for Magnetopause-to-Aurora Global Exploration
IMAS	Integrated Multispectral Atmospheric Sounder
IMEX	Inner Magnetosphere Explorer
IMP	Interplanetary Monitoring Platform
INTEGRAL	International Gamma-Ray Astrophysics Laboratory
IPO	Integrated Program Office
ISEE	International Sun-Earth Explorer
ISM	interstellar medium
ISTP	International Solar-Terrestrial Physics program
JPL	Jet Propulsion Laboratory
LISA	Laser Interferometer Space Antenna

MAP	Microwave Anisotropy Probe
MCO	Mars Climate Observer
MIDEX	Medium Explorer (program)
MMS	Magnetospheric Multi Scale
MO&DA	mission operations and data analysis
MoO	mission of opportunity
MPL	Mars Polar Lander
NASA	National Aeronautics and Space Administration
NEAR	Near Earth Asteroid Rendezvous
NGST	Next Generation Space Telescope
NMS	Neutral Mass Spectrometer
NOAA	National Oceanic and Atmospheric Administration
NPOESS	National Polar-Orbiting Operational Environmental Satellite System
NPP	NPOESS Preparatory Project
NRC	National Research Council
NSTC	National Science and Technology Council
OSS	Office of Space Science (NASA)
PI	principal investigator
PIDDP	Planetary Instrument Definition and Development Program
POES	Polar-orbiting Operational Environmental Satellite
QuikSCAT	Quick Scatterometer
R&A	research and analysis
ROSAT	Roentgen Satellite
RXTE	Rossi X-ray Timing Explorer
SAMPEX	Solar Anomalous and Magnetospheric Particle Explorer
SAR	synthetic aperture radar
SAX	Satellite per Astronomìa in Raggi X
SDI	Strategic Defense Initiative
SeaWiFS	Sea Viewing Wide Field of View Sensor
SEC	Sun-Earth Connection
SIM	Space Interferometry Mission
SIRTF	Space Infrared Telescope Facility
SME	Solar Mesosphere Explorer
SMEX	Small Explorer (program)
SNOE	Student Nitric Oxide Experiment (program)
SOFIA	Stratospheric Observatory for Infrared Astronomy
SOHO	Solar and Heliospheric Observatory
SSB	Space Studies Board
SSTI	Small Spacecraft Technology Initiative
STD	science and technology definition
STEDI	Student Explorer Demonstration Initiative
STEREO	Solar Terrestrial Relations Observatory
STP	Solar-Terrestrial Probe
STScI	Space Telescope Science Institute

SWAS	Submillimeter Wave Astronomy Satellite
TERRIERS	Tomographic Experiment Using Radiative Recombinative Ionospheric Extreme Ultraviolet and Radio Sources
TIMED	Thermosphere-Ionosphere-Mesosphere Energetics and Dynamics
TOMS	Total Ozone Mapping Spectrometer
TOPEX	Ocean Topography Experiment
TPF	Terrestrial Planet Finder
TRACE	Transition Region and Coronal Explorer
TRMM	Tropical Rainfall Mapping Mission
UARS	Upper Atmosphere Research Satellite
UNEX	University Class Explorer (program)
USGCRP	U.S. Global Change Research Program
USRA	Universities Space Research Association
VLA	Very Large Array
WIRE	Wide-field InfraRed Explorer
XEUS	X-ray Evolving Universe Spectroscopy mission
XMM	X-ray Multi-Mirror mission

G

Biographies of Committee Members

Daniel N. Baker, chair of the committee, is primarily interested in research into plasma physical and energetic particle phenomena in the planetary magnetospheres and in the Earth's magnetosphere, and he also conducts research in space instrument design, space physics data analysis, and magnetospheric modeling. He was a research associate at the University of Iowa's Department of Physics from 1974 to 1975 and a research fellow at the California Institute of Technology from 1975 to 1977. In 1977, he joined the physics research staff at Los Alamos National Laboratory, and he became leader of the Space Plasma Physics Group in 1981. From 1987 to 1994, he was the chief of the Laboratory for Extraterrestrial Physics at NASA Goddard Space Flight Center. Dr. Baker is currently the director of the Laboratory for Atmospheric and Space Physics and professor of astrophysical and planetary sciences at the University of Colorado. He has served on several National Research Council (NRC) and National Aeronautics and Space Administration (NASA) committees.

Fran Bagenal is currently the associate chair of the Department of Astrophysical and Planetary Sciences at the University of Colorado at Boulder. Her research interests include the synthesis of data analysis and theory in the study of space plasmas. She specializes in the field of planetary magnetospheres, particularly Jovian magnetospheres, and solar corona. Dr. Bagenal received NASA Group Achievement Awards in 1981, 1986, 1990, and 1996. She is currently a member of the National Research Council's Space Studies Board and the Committee on International Space Programs. She is also an interdisciplinary scientist for the Galileo Project, a coinvestigator for the Voyager Plasma Science Experiment of the Planetary Exploration Division of NASA, and an associate of the Laboratory for Space and Atmospheric Physics at the University of Colorado at Boulder. In addition, Dr. Bagenal is a member of the American Astronomical Society, the American Geophysical Union, the American Physical Society, the Royal Astronomical Society, and the American Association of Physics Teachers.

Robert L. Carovillano is a member of the Boston College faculty. In space physics research, Dr. Carovillano has published on a broad spectrum of topics in pure theory and data analysis, including magnetospheric energy theorems and related topics. He has served on national advisory committees of the National Academy of Sciences, the National Center for Atmospheric Research, NASA, and the National Science Foundation (NSF) and has chaired several such advisory committees. Dr. Carovillano has been a principal investigator on many research grants and contracts funded by the NSF, NASA, the Office of Naval Research, and the Air Force. He was a visiting senior scientist at NASA Headquarters Office of Space Science. At NASA he was responsible for the supervision of several programs and research initiatives in space physics but was most deeply engaged in optimizing mission scientific accomplishments and opportunities.

Richard G. Kron is a professor in the Department of Astronomy and Astrophysics at the University of Chicago, director of the Yerkes Observatory, and head of the Experimental Astrophysics Group, Fermilab National Accelerator Laboratory. His research interests include optical studies of galaxies. Dr. Kron's primary responsibility for the Experimental Astrophysics Group is data system development for the Sloan Digital Sky Survey. This group is also responsible for building a drift-scan charge coupled device camera that was commissioned at the Yerkes Observatory and then deployed to the ARC 3.5-m telescope at Apache Point Observatory. Dr. Kron's prior NRC service includes membership on the Steering Committee for the Task Group on Space Astronomy and Astrophysics and the Panel on Cosmology.

George A. Paulikas has been at the forefront of advances in space science and space systems, making innumerable technical contributions to national security space systems. He retired after 37 years at the Aerospace Corporation, having joined Aerospace in 1961 as a member of the technical staff, later becoming department head, laboratory director, vice president, and senior vice president. He became executive vice president in 1992. He received the company's highest award, the Trustees' Distinguished Achievement Award, in 1981 in recognition of research leading to a new understanding of the dynamics of space radiation and its effect on spacecraft. Dr. Paulikas's other awards and honors include the Jimmy Doolittle Fellowship Award, the National Reconnaissance Office Gold Medal, the Air Force Space Division Award for Excellence, and the Air Force Meritorious Civilian Service medal, both in 1981 and 1996.

R. Keith Raney, Jr., is principal professional staff scientist with the Johns Hopkins University Applied Physics Laboratory (APL). He is on special assignment in the APL Space Department, where his responsibilities are for new initiatives in microwave remote sensing and satellite system configurations. Prior to his employment at Johns Hopkins, Dr. Raney was at the Canada Centre for Remote Sensing, where he was chief radar scientist and cofounder of RADARSAT, Canada's first remote sensing satellite program. He participated in the conceptual design phase of several satellite programs.

Pedro L. Rustan, Jr., is retired from the U.S. Air Force. He is experienced in the management of design and integration of advanced technologies. As director for Sensor Integration at the Ballistic Missile Defense Organization (BMDO) from 1989 to 1991, Col. Rustan managed the SPEAR and POAM programs. From 1991 to 1994, Col. Rustan was mission director for BMDO's Clementine program from inception to full operational capability. He was the director of Small Satellite Development at the National Reconnaissance Office from 1994 to early 1997, where he was responsible for conceptualizing a constellation of smaller imaging spacecraft. Currently, Col. Rustan is a consultant. Among his most recent honors are the NASA Outstanding Leadership Medal (1994), the Goddard Space Flight Center Nelson Jackson Award (1995), and the National Reconnaissance Office Medal of Superior Service (1996, 1997). Col. Rustan is a member of the Space Studies Board's executive committee.